Theoretical perspectives on signal transduction pathways depending on regulated proteolysis

D i s s e r t a t i o n

zur Erlangung des akademischen Grades
d o c t o r   r e r u m   n a t u r a l i u m
(Dr. rer. nat.)
im Fach Biophysik

eingereicht an der
Mathematisch-Naturwissenschaftlichen Fakultät I
der Humboldt-Universität zu Berlin

von
Diplom-Biophysikerin Bente Kofahl
geboren am 22.02.1978 in Berlin

Präsident der Humboldt-Universität zu Berlin:
Prof. Dr. Jan-Hendrik Olbertz

Dekan der Mathematisch-Naturwissenschaftlichen Fakultät I:
Prof. Dr. Andreas Herrmann

Gutachter/innen:
1. Prof. Dr. Edda Klipp
2. Prof. Dr. Thomas Höfer
3. Prof. Dr. Nils Blüthgen

Tag der mündlichen Prüfung: 13. Juli 2012

Bibliografische Information der Deutschen Nationalbibliothek

Die Deutsche Nationalbibliothek verzeichnet diese Publikation in der
Deutschen Nationalbibliografie; detaillierte bibliografische Daten sind
im Internet über http://dnb.d-nb.de abrufbar.

ISBN 978-3-8325-3337-3

Logos Verlag Berlin GmbH
Comeniushof, Gubener Str. 47,
10243 Berlin
Tel.: +49 (0)30 42 85 10 90
Fax: +49 (0)30 42 85 10 92
INTERNET: http://www.logos-verlag.de

Zusammenfassung

In eukaryotischen Zellen wird die Aufnahme und Weiterleitung extrazellulärer Reize durch Signaltransduktionswege (Signalwege) vermittelt. Auch wenn eine Vielzahl verschiedener Signaltransduktionswege existiert, gibt es einige allgemeine Mechanismen, die wiederholt auftreten. Eines dieser Regulationsprinzipien ist die Kontrolle der Signalweiterleitung über die regulierte Proteolyse. Dabei werden in unstimulierten Zellen Schlüsselproteine des Signaltransduktionsweges kontinuierlich abgebaut, um die Signalweiterleitung zu verhindern. Trifft ein äußerer Reiz auf die Zelle, wird der Abbau verringert und somit Weiterleitung des Signals ermöglicht.

Zwei prominente Signalwege, die diesen Mechanismus nutzen, sind i) der Wnt/β-Catenin- und ii) der nicht-kanonische NF-κB- (*nuclear factor κ-light-chain-enhancer of activated B cells*) Signalweg. Im Wnt/β-Catenin-Signalweg ist das Schlüsselprotein der Transkriptionsfaktor β-Catenin, der in unstimulierten Zellen kontinuierlich über einen streng regulierten Mechanismus abgebaut wird. Der nicht-kanonische NF-κB-Signalweg resultiert in der Bildung von p52/RelB-Komplexen. Um diese bilden zu können, wird die Kinase NIK (*NF-κB inducing kinase*) benötigt, deren Konzentration über die regulierte Proteolyse kontrolliert wird. In der vorliegenden Arbeit werden Methoden der mathematischen Modellierung genutzt, um diese beiden Signaltransduktionswege zu untersuchen und zu deren besserem Verständnis beizutragen.

Die Grundlage für die Analysen des Wnt/β-Catenin-Signalweges ist das mathematische Modell dieses Signalweges, welches von Lee und Kollegen basierend auf experimentellen Untersuchungen in Zellextrakten entwickelt wurde (Lee et al., 2003). Dieses Modell wird zur Behandlung unterschiedlicher Fragestellungen erweitert und analysiert. Es wird untersucht, welche Effekte durch Hemmungen einzelner oder zweier Signalwegreaktionen hervorgerufen werden. Dabei werden sowohl Auswirkungen auf die β-Catenin-Konzentration im stationären Zustand (stationäre β-Catenin-Konzentration) als auch Änderungen im Verhältnis der β-Catenin-Konzentrationen im stimulierten und nicht-stimulierten stationären Zustand (induzierbares β-Catenin-Verhältnis) untersucht. Es zeigt sich, dass diese beiden Charakteristika unterschiedlich durch Inhibitionen beeinflusst werden. In verschiedenen Krankheiten, insbesondere unterschiedlichen Arten von Krebs, wurde eine Fehlregulation des

Wnt/β-Catenin-Signalweges entdeckt, die auf Mutationen beteiligter Proteine zurückgeführt wird. Unter solch krankhaften Bedingungen ist die Wirkung der Reaktionshemmungen von besonderem Interesse, da Substanzen wie Inhibitoren eingesetzt werden, um der Erkrankung entgegenzuwirken. Ein Protein des Signalweges, das häufig Mutationen zeigt, ist APC (*adenomatous polyposis coli*). Durch die APC-Mutation verändert sich der Einfluss, der durch Hemmungen von Reaktionen bewirkt werden kann. Die Untersuchung zeigt, dass es Inhibitionen gibt, die in gesunden Zellen einen starken Einfluss ausüben, in abnormen Zellen jedoch keinen oder nur einen schwachen Effekt erzielen können. Zur Entwicklung wirkungsvoller Medikamente ist die Kenntnis solcher Verschiebungen von Wirkungseffektivitäten von großem Interesse. Die Analysen unterstreichen weiterhin die wichtige Rolle des alternativen β-Catenin-Abbaus und die Rolle der positiven Regulation der β-Catenin-Konzentration durch den APC/β-Catenin-Komplex.

Ein weiteres Protein, das in Tumoren häufig mutiert vorliegt, ist β-Catenin. Ein mathematisches Modell, das sowohl Wildtyp- als auch mutiertes β-Catenin berücksichtigt, wird entwickelt und analysiert, um die Effekte der Koexistenz der beiden Proteinformen zu untersuchen. Es zeigt sich, dass die Dynamiken der beiden β-Catenin-Formen einander beeinflussen. Durch die Anwesenheit von mutiertem β-Catenin steigt die stationäre Konzentration von Wildtyp-β-Catenin, da beide β-Catenin-Formen um die Bindungspartner konkurrieren. Der Haupteffekt wird dabei durch das Binden von APC vermittelt. Theoretische Vorhersagen werden durch experimentelle Untersuchungen bestätigt. Für zukünftige Analysen der Effekte von mutiertem β-Catenin auf die Genexpression wird ein Minimalmodell entwickelt, das mutiertes und Wildtyp-β-Catenin berücksichtigt. Dieses kann teilweise analytisch untersucht werden und zeigt die gleichen Signalcharakteristika für Wildtyp- und mutiertes β-Catenin wie das detaillierte Modell.

Zeitaufgelöste Konzentrationsmessungen von β-Catenin in zwei Zelltypen zeigen unterschiedliche Dynamiken. Um diese erklären zu können, wird das mathematische Modell um Prozesse erweitert, die in Zellen wichtige Funktionen haben, im Zellextrakt jedoch nicht auftreten und somit im Modell von Lee und Kollegen nicht integriert sind. Dies sind ein transkriptioneller Regulationsmechanismus sowie die Möglichkeit von β-Catenin, in Prozesse der Zelladhäsion involviert zu sein. Die beiden Modellerweiterungen haben unterschiedliche Auswirkungen auf das Systemverhalten. Das zelluläre mathematische Modell ist in der Lage, die unterschiedlichen Dynamiken von β-Catenin in den verschiedenen Zelltypen zu reproduzieren. Zudem zeigt das zelluläre mathematische Modell, dass Hemmungen unterschiedlich starke Effekte vermitteln können in Abhängigkeit davon, welcher konkrete Zelltyp durch das zelluläre Modell repräsentiert wird. Auch die Effekte, die durch APC-Mutationen hervorgerufen werden, sind unterschiedlich stark ausgeprägt, je nachdem welches zelluläre mathematische Modell zur Untersuchung genutzt wird. Die theoretischen Analysen betonen die Wichtigkeit zelltypspezifischer Informationen, um entsprechende zelltypspezifische Vorhersagen machen zu können.

Im Gegensatz zum Wnt/β-Catenin-Signalweg gibt es bislang kein mathematisches Modell des nicht-kanonischen NF-κB-Signalweges. Bisherige Modelle konzentrieren sich auf den kanonischen NF-κB-Signalweg. In der vorliegenden Arbeit wird ein Modell für den nicht-kanonischen NF-κB-Signaltransduktionsweg entwickelt, das die molekularen Prozesse zentraler Komponenten des Signalweges beschreibt. Es beruht auf einer Literaturrecherche, der Expertise von Kooperationspartnern sowie experimentellen Beobachtungen aus der Kooperationsgruppe. Das entwickelte Modell kann die gemessenen Dynamiken verschiedener Komponenten des Signalweges widerspiegeln. Darüber hinaus kann es die Effekte zahlreicher Mutationen von Signalwegskomponenten erklären. Diese Untersuchung unterstreicht, dass das Zusammenspiel der einzelnen Komponenten durch das mathematische Modell treffend beschrieben wird. Abweichungen zwischen Modell und Experiment können damit erklärt werden, dass einerseits der nicht-kanonische NF-κB-Signalweg unabhängig vom kanonischen Signalweg betrachtet wird und andererseits die Regulation der NIK-Konzentration vereinfacht dargestellt wird. Durch ein detailliertes mathematisches Modell dieser Regulation kann die Abweichung zwischen Modell und Experiment aufgehoben werden. Das mathematische Modell des nicht-kanonischen NF-κB-Signalweges erlaubt weiterhin Vorhersagen von Effekten in Störungsexperimenten. Solche Experimente werden zurzeit durchgeführt, um das mathematische Modell weiter zu verifizieren.

Insgesamt zeigt die vorliegende Arbeit, dass Methoden der mathematischen Modellierung zu einem tieferen Verständnis der untersuchten Signaltransduktionswege beitragen.

Contents

1 Introduction

The ability of cells to detect and to respond to external stimuli is of high importance. Unicellular organisms use this ability to detect nutrients, to adapt to changing osmotic conditions, to change between particular metabolic pathways, sense light or temperature or to find mating partners. In multi-cellular organisms, cells may communicate with each other, influencing each other's behaviour, and may react to changes in their environment by exchanging signals. The capability of cells to correctly respond to their environment is the basis of development, tissue repair and immunity. Depending on the kind of stimulation, cells can start to grow, divide, move, differentiate or die. This is possible as the stimuli lead to an altered expression of specific target genes and/or changes in the metabolism. The response to a particular stimulus or a combination of several stimuli is cell-type-specific, meaning that a stimulus that leads to a specific effect in one cell-type may cause other cellular behaviours in other cell-types. Like the cellular answers the signals show a wide variety as well. These signalling molecules could for instance be hormones, growth factors, neurotransmitters or cytokines.

To respond to the stimulus the signal has to be transmitted from the outer cell surface to the intercellular target. Therefore, specific signal transduction pathways (signalling pathways) evolve, forming a complex intracellular network. Often, a stimulus reaches the cell surface and alters the state of a receptor, transmitting the signal into the cell. It is also possible that a small molecule enters the cell directly. Several intracellular signalling molecules work together to transfer the signal to reach its targets. To accomplish that, coordinated protein-protein interactions are necessary. Proteins may be recruited, bind, dissociate, change conformations or modify each other, leading to the transduction of the signal. Furthermore, the participating molecules may change their location and number. Often, the signal is amplified along the signalling pathway as one protein may activate multiple effector proteins. The signal transduction leads to the localisation and regulation of transcription factors that control and modulate the expression of target genes. Their products are necessary to generate the appropriate cellular response.

Besides mechanisms for the signal transmission there are mechanisms for downregulation and termination of the signal. Amongst other effects this provides the cell with the ability to respond to a further stimulus.

The mechanism of regulated proteolysis

Although a variety of signalling pathways emerge, some general signalling mechanisms occur repeatedly. One common principle is the regulation of a pathway by regulated proteolysis. In this case, a stimulus does not directly lead to the activation of the key molecule but alters its availability in the cell. This could be accomplished by i) the direct regulation of its production or its degradation or ii) by the regulation of the concentration of an essential activator or inhibitor. In cases i), the concentration of the key molecule is controlled directly, while in cases ii) its usable form is controlled indirectly by regulating the concentration of modifiers. Well-described examples of signalling pathways that depend on regulated proteolysis are the Wnt/β-catenin signalling pathway and the canonical as well as the non-canonical NF-κB (nuclear factor κ-light-chain-enhancer of activated B cells) signalling pathways.

While in the Wnt/β-catenin pathway the concentration of the key molecule β-catenin itself is constrained by regulated proteolysis, in both NF-κB signalling branches, the concentration of essential modifiers is strictly controlled by regulated proteolysis. In the canonical branch an inhibitor is degraded in response to a stimulus. In contrast, in the non-canonical signalling branch the degradation of an activator is blocked upon pathway activation. Although they differ in this general feature these signalling pathways all belong to the family of pathways that are dependent on the regulated proteolysis. This thesis concentrates on the Wnt/β-catenin pathway and on non-canonical NF-κB signalling.

Modelling signal transduction pathways

For years, specific signalling pathways as well as general signalling features and design principles have been the object of intensive experimental and theoretical research. Nevertheless, the functioning and regulation of signalling pathways are still not well understood. Many diseases, as for instance various types of cancer or autoimmune disease, are linked to dysregulated signal transduction pathways. Hence, the knowledge of the pathway's functioning would be of great help for the understanding of these diseases and the development of effective intervention strategies.

The precise mathematical description of a specific signalling pathway may contribute to the understanding of the pathway function, regulatory properties and underlying mechanisms of specific features observed (e.g. (Becker et al., 2010; Kiel and Serrano, 2009; Klipp and Liebermeister, 2006; Klipp et al., 2005; Kofahl and Klipp, 2004; Lee et al., 2003; Schilling et al., 2009; Schoeberl et al., 2002)). Modelling may help to distinguish between possible pathway topologies. The theoretical reflection of experimental data helps to interpret experimental observations and to suggest meaningful further experiments. New hypotheses could be generated and the most informative experiments to test them can be suggested. Additionally, predictions for effective interventions of the pathway are possible. This in turn gives information about drug effects or consequences of mutations on cellular processes. Modelling crosstalks between several pathways gives insights into their interplay and how the interference of one pathway influences the functioning of the connected one.

The theoretical analysis of general mechanisms and design principles helps to understand the interrelation between a system's design and the system's behaviour (Alon, 2007; Blüthgen et al., 2009; Ferrell, 2002; Fritsche-Guenther et al., 2011; Salazar and Höfer, 2009; Tyson et al., 2003). It contributes to the characterisation of signalling properties, e.g. whether a network's structure is able to generate oscillatory or bistable behaviour (Huang and Ferrell, 1996; Kholodenko, 2000). Specific structures may provide adaptation to or switches between transient and sustained responses. Furthermore, general dependencies of the pathway's readout on the signal strength or duration have been analysed, both for isolated pathways and crosstalks between pathways (Behar et al., 2007a; Behar et al., 2007b; Haney et al., 2010). With regard to crosstalk, effects of various pathway wirings have been investigated and distinct measures to characterise crosstalk have been introduced (Haney et al., 2010; Komarova et al., 2005; Schaber et al., 2006; Seaton and Krishnan, 2011). Conclusions have been drawn about the sensitivity or robustness of a general mechanism. General signalling characteristics such as signalling time or signal duration have been defined (Heinrich et al., 2002; Llorens et al., 1999). Analysing how they depend on pathway components has revealed new insights into a pathway's functioning (Heinrich et al., 2002; Hornberg et al., 2005). Conditions to amplify or dampen a signal have been identified (Binder and Heinrich, 2004; Heinrich et al., 2002).

Objectives

In this thesis, two signalling pathways depending on the mechanisms of regulated proteolysis are investigated. They have been objects of research for years. However, it is still unclear what the advantages of the signal transduction via this mechanism are. At first glance, this principle of regulation seems to be relatively uneconomical by dissipating energy in a resting cell due to the constant production of proteins that are immediately degraded thereafter. Signal transduction via transient modifications of proteins that are available in the cell appears to be less energy consuming. This mechanism is for instance realised in the mitogen-activated protein kinase (MAP kinase; MAPK) cascade. However, the energy needed is possibly not the deciding criterion in regard to whether a signalling pathway is established or not. Other criteria, such as a fast response to stimuli or controlled protein levels, may be more crucial. Signalling pathways that are based on regulated proteolysis are involved especially in developmental processes. Maybe the basis of this regulatory type derives its origin from the special requirements during development.

A first step to investigate the advantages of the regulated proteolysis is an understanding of the pathways depending on this general mechanism of regulation. Their analyses may contribute to elucidate the gains of the signalling mechanism. In this work, mathematical methods are used to get deeper insights in the functioning of pathways that use the mechanism of the regulated proteolysis. Mathematical models of the canonical Wnt/β-catenin pathway and the non-canonical NF-κB signalling pathway are developed and analysed.

In this thesis, the priority is the deeper understanding of the functioning of these pathways themselves. For the canonical Wnt/β-catenin signalling pathway a mathematical model exists that is based on the molecular processes (Lee et al., 2003). This model is further analysed and extended to address distinct questions. Several diseases, such as various kinds of cancer, are regarded to be connected with a dysfunction of this signalling pathway. For an effective treatment, a deep understanding of the signalling pathway is necessary, both in healthy and in abnormal cells. This may permit predictions about effective interventions and drug effects. As distinct mutations occur exclusively in specific tissues, cell-type-specific properties seems to be of major importance. Regarding the Wnt/β-catenin pathway the following questions are asked:

1) How does the pathway operate in healthy and in abnormal cases? In particular, what are the effects of mutations of various pathway components on distinct signalling properties?

2) Which is the best drug target which will influence the pathway significantly in the diseased situation but only slightly under healthy circumstances?

3) Are there any differences in the pathway's function in different cell-types? Does a treatment affect all cell-types similarly?

In section 2, the main part of this thesis, the Wnt/β-catenin pathway is analysed. The pathway is introduced in section 2.1. A biological overview as well as a review about modelling successes of this pathway is provided. The analyses of the pathway focus on the effects of mutations on the pathway's function (sections 2.2 and 2.3) and cell-type-specific features (section 2.4).

Concerning the non-canonical NF-κB signal transduction pathway, this work is focussed on the development of a mathematical model for that pathway. A biological overview of non-canonical NF-κB signalling is provided in section 3.1. Section 3.2 deals with the development of a mathematical model of the non-canonical NF-κB-signalling branch and its initial analyses.

2 The canonical Wnt/β-catenin signalling pathway

The Wnt/β-catenin pathway regulates the concentration of β-catenin and is critically involved in processes of development, organogenesis, proliferation, regeneration and cell-cell adhesion. Components of this signalling pathway have been found to be mutated in various diseases, most notably in cancer (Aulehla and Pourquié, 2008; Bienz, 2005; Cadigan and Peifer, 2009; Clevers, 2006; Dequeant and Pourquie, 2008; Klaus and Birchmeier, 2008; Logan and Nusse, 2004; MacDonald et al., 2009; Moon et al., 2004; Polakis, 2007).

Section 2.1 introduces this signalling pathway, provides a biological overview and discusses existing modelling approaches. Sections 2.2, 2.3 and 2.4 deal with our analyses of the Wnt/β-catenin signalling pathway. Since the pathway is dysregulated in several types of cancer, the effects of specific mutations on the pathway function are of particular interest. Understanding their impact will help to develop medical strategies. The theoretical analysis of the effects of single inhibitors and inhibitor combinations allows the prediction of optimal intervention points and optimal drug combinations in the presence and absence of different mutations. Mutations may have different effects in different cell-types and particular mutations may occur exclusively in certain tissues. Therefore, cell-type-specificity plays a decisive role for signal transduction as well as for finding effective drugs.

2.1 Introduction of canonical Wnt/β-catenin signalling[1]

2.1.1 Biological background

β-Catenin is a bifunctional protein that plays a prominent role in cell-cell adhesion by binding to E-cadherin and α-catenin as well as in transcriptional regulation, reviewed in (Heuberger and Birchmeier, 2010). The protein is constantly synthesised and its degradation is highly regulated. A prerequisite for the degradation of β-catenin is its sequential phosphorylation. Following a priming phosphorylation at serine Ser^{45} mediated by the casein kinase1α (CK1α)

---

[1] Parts of this section have been published in Kofahl, B. and Wolf, J. (2010). Mathematical modelling of Wnt/beta-catenin signalling. *Biochem Soc Trans* **38**, 1281-1285.

(Amit et al., 2002; Liu et al., 2002), β-catenin is phosphorylated at three further amino-terminal serine and threonine residues (Ser[33], Ser[37] and Thr[41] in humans) by the glycogen synthase kinase-3 (GSK3) (Ikeda et al., 1998). These modifications take place within the so-called destruction complex. Besides GSK3 and CK1α, the complex contains additional proteins such as phosphatases and the two scaffold proteins APC (adenomatous polyposis coli) and Axin-1 (Kimelman and Xu, 2006; Xing et al., 2003). Phosphorylated β-catenin is subsequently recognised by β-TrCP (β-transducin-repeat-containing protein), a subunit of the ubiquitin ligase complex, and marked with ubiquitin. This leads to its proteasome-dependent degradation (Aberle et al., 1997; Liu et al., 1999). The exact composition of the destruction complex is still not resolved. Neither all participating proteins nor their stoichiometry are known.

The application of a Wnt stimulus results in the dissociation of the destruction complex. As a consequence, less β-catenin is degraded and the protein enters the nucleus. How β-catenin is translocated into and out of the nucleus is not completely understood and still under experimental investigation (Henderson and Fagotto, 2002; Krieghoff et al., 2006; Städeli et al., 2006). Nuclear β-catenin interacts with HMG (high mobility group) box transcription factors of the TCF/LEF protein family (T-cell-factor/lymphocyte enhance factor) and further co-factors regulating the transcription of various target genes, for example myc, cyclin D1, siamois and FGF4 (Brannon et al., 1997; Daniels and Weis, 2005; Hecht and Stemmler, 2003; Hecht et al., 2000; Mosimann et al., 2009; Riese et al., 1997). Target genes also encode for components and regulators of the Wnt/β-catenin pathway itself, such as Dickkopf (Dkk), β-TrCP, LEF and Axin-2 (Hovanes et al., 2001; Jho et al., 2002; Leung et al., 2002; Lustig et al., 2002; Niida et al., 2004; Spiegelman et al., 2000). Axin-2 is a negative regulator of the signalling pathway as it is functionally equivalent to Axin-1 and may fulfil the same functions. Axin-1 and Axin-2 are denoted as Axin in the modelling context. Dickkopf acts as an inhibitor of the pathway by interacting with the co-receptor LRP5/6 (low-density lipoprotein receptor-related protein-5 and -6).

How the stimulus induces the decomposition of the destruction complex is still object of intensive research (Angers and Moon, 2009; Cadigan and Liu, 2006; He et al., 2004; Huang and He, 2008; Tauriello et al., 2010). A Wnt stimulus directs the receptor Frizzled and the co-receptor LRP5/6 into close proximity (Bilic et al., 2007; Cong et al., 2004; Niehrs and Shen, 2010). Subsequently, several events can take place, including the phosphorylation of LRP5/6 by CK1 and GSK3, recruitment of proteins participating in the destruction complex, e.g. Axin and GSK3, and the phosphorylation of the protein Dishevelled (Dsh) (Davidson et al., 2005; Mao et al., 2001; Takada et al., 2005; Tamai et al., 2004; Wu et al., 2009; Yanagawa et al., 1995; Zeng et al., 2005). Dsh is involved in the recruitment of proteins as well as in the regulation of the receptors within the pathway (Zeng et al., 2008). Recent experiments have revealed the importance of the ubiquitination state of Dsh for its ability to polymerise and transduce the signal (Tauriello et al., 2010). The recruitment of proteins to the plasma membrane and the related modifications lead to an inhibition of the destruction complex, and hence, to a reduced degradation of β-catenin. A schematic description of the pathway is given in Figure 2.1.

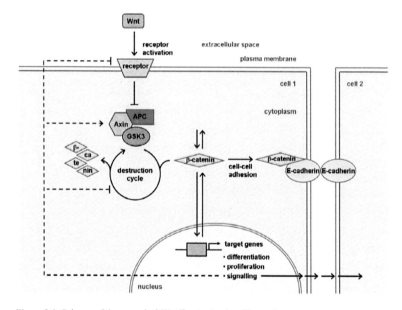

Figure 2.1: Scheme of the canonical Wnt/β-catenin signalling pathway.
Without a Wnt stimulus, β-catenin is constantly degraded by the destruction complex. This complex contains APC, Axin and GSK3. A Wnt stimulus leads to its dissociation. Hence, less β-catenin is degraded. The protein can translocate into the nucleus regulating the expression of various target genes, including components or regulators of the Wnt/β-catenin pathway itself. These regulations act on different levels, indicated by broken lines. β-Catenin is also able to bind E-cadherin participating in cell-cell adhesion processes.

Besides these canonical processes, crosstalks with non-canonical Wnt pathways, such as the Wnt/Ca^{2+} and the planar cell polarity (PCP) pathway, have been described (Kestler and Kuhl, 2008; McNeill and Woodgett, 2010; van Amerongen et al., 2008) as well as interactions with other signalling pathways, such as EGF (epidermal growth factor), NF-κB, BMP (bone morphogenetic protein), Hedgehog, Notch or PI3K (phosphoinositide 3-kinase) signalling (Brabletz et al., 2009; Cho et al., 2008; Hayward et al., 2005; Ille et al., 2007; Ji et al., 2009; Nusse, 2003; Saegusa et al., 2007; Villacorte et al., 2010; Wang et al., 2004).

2.1.2   Mathematical modelling of the canonical Wnt/β-catenin pathway

In the last years, quantitative and qualitative mathematical modelling approaches have been used to describe this signalling pathway and have provided deeper insights in its functioning in time and space. Here, an overview of these models is given.

The first theoretical study on this signalling pathway has been a quantitative model for canonical Wnt/β-catenin signalling based on measurements in *Xenopus* oocyte extracts (Lee et al., 2003). This kinetic model describes the molecular reactions of the pathway's core components. Numerous modifications and further analyses with respect to signalling characteristics, the interplay of its components and feedback loops have been performed (Cho et al., 2006; Goentoro and Kirschner, 2009; Krüger and Heinrich, 2004; Mirams et al., 2010; Wawra et al., 2007). β-Catenin does not only act as a transcriptional regulator, but is also involved in cell-cell adhesion. This bifunctionality of β-catenin and/or spatial effects of signalling have been included in various models (Murray et al., 2010; Ramis-Conde et al., 2008; Sick et al., 2006; van Leeuwen et al., 2009; van Leeuwen et al., 2007). The canonical Wnt pathway alone (Jensen et al., 2010) as well as its crosstalk with other signalling pathways have been analysed in the context of development and tumourigenesis (Goldbeter and Pourquie, 2008; Kim et al., 2007; Rodriguez-Gonzalez et al., 2007; Shin et al., 2010). A time line of the models is shown in Figure 2.2.

In other studies, mathematical models focussing on the intra- and intercellular regulatory interactions on the level of gene expression networks have been generated. They neglect the details of the signalling pathways and have been developed to explain the expression patterns of the segment polarity genes along several cells (Albert and Othmer, 2003; Chaves et al., 2005; von Dassow et al., 2000). As the present work focuses on the signalling pathway, these models are not further considered.

Figure 2.2: Time line of the mathematical models of the Wnt/β-catenin signalling pathway.
The models are indicated by rectangles with the name of the first author and placed above the
year published. A white background indicates models concentrating on segmentation in early
development. Models focussing on mechanisms of the intracellular signal transduction are
highlighted with a grey background. Models taking both aspects into account are marked both
white and grey. Black rectangles represent models focussing on spatial signalling aspects
within tissues.
The references are: Lee (Lee et al., 2003), Krüger (Krüger and Heinrich, 2004), Sick (Sick et
al., 2006), Cho (Cho et al., 2006), Rodriguez-Gonzales (Rodriguez-Gonzalez et al., 2007),
van Leeuwen (van Leeuwen et al., 2007), Kim (Kim et al., 2007), Wawra (Wawra et al.,
2007), Ramis-Conde (Ramis-Conde et al., 2008), Goldbeter (Goldbeter and Pourquie, 2008),
Goentoro (Goentoro and Kirschner, 2009), van Leeuwen (van Leeuwen et al., 2009), Jensen
(Jensen et al., 2010), Murray (Murray et al., 2010), Shin (Shin et al., 2010), Mirams (Mirams
et al., 2010).

Insights by the quantitative kinetic core-model of the pathway

In a combined experimental and theoretical approach, the groups of Marc W. Kirschner and
Reinhart Heinrich have developed a mathematical model of the canonical Wnt/β-catenin
signal transduction pathway (Lee et al., 2003). This model forms the basis of the analyses in
this thesis. It is based on comprehensive data measured in *Xenopus* oocyte extracts including
concentrations, fluxes and characteristic times. In the model, the interactions of the core
components β-catenin, Axin, APC, GSK3, TCF and Dsh are taken into account. A scheme is
shown in Figure 2.3. The individual reactions and components are described in the legend.

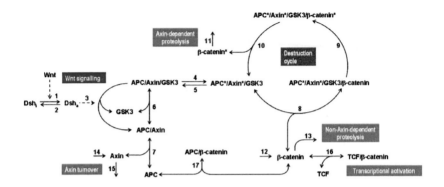

Figure 2.3: Reaction scheme of the core-model of the Wnt/β-catenin pathway.

The model takes into account the core components APC, Axin, β-catenin, GSK3, Dsh and TCF. β-Catenin is produced and degraded via a phosphorylation-independent mechanism (reactions 12 and 13, respectively). It reversibly binds to the destruction complex APC*/Axin*/GSK3 (reaction 8) and is phosphorylated within that complex (reaction 9). After dissolution from the complex (reaction 10), phosphorylated β-catenin is degraded (reaction 11). TCF (reaction 16) and APC (reaction 17) reversibly bind β-catenin. APC also binds Axin (reaction 7) that is produced and degraded via the reactions 14 and 15, respectively. GSK3 reversibly binds the APC/Axin complex (reaction 6) and leads to a subsequent phosphorylation of both components (reaction 4). Phosphatases have the ability to dephosphorylate them again (reaction 5). A Wnt stimulus activates inactive Dsh (Dsh$_i$) (reaction 1), which is able to stimulate the dissociation of GSK3 from the complex APC/Axin/GSK3 (reaction 3). Deactivation of active Dsh (Dsh$_a$) is included via reaction 2. Components in a complex are separated by a slash. The phosphorylation of a component is indicated by an asterisk. The numbers next to arrows denote the number of the particular reaction. One-headed arrows denote reactions taking place in the indicated direction. Dashed arrows represent activation steps and double-headed arrows illustrate binding reactions.

The system is described by a set of 15 coupled ordinary differential equations (ODEs) with 31 parameters. With the knowledge about concentrations and assumptions about time scales, several differential equations have been replaced by algebraic equations (Krüger and Heinrich, 2004; Lee et al., 2003). The reduced model contains seven ODEs, eight algebraic equations and 19 parameters. It is referred to as core-model in the following.

Additional experiments have been used to validate the model. The subsequent analysis has led to numerous new insights into the functioning of the pathway, see also (Tolwinski and Wieschaus, 2004). A crucial observation has been the low abundance of Axin. The fast Axin turnover has a strong effect on the steady state, amplitude and duration of the β-catenin signal, leading to the conclusion that the regulation of the Axin turnover could be an effective intervention target for drugs. This has been emphasised by a sensitivity analysis of the system, indicating whether processes or components have oncogenic or tumour suppressor effects. The prediction has been confirmed recently by the finding of new inhibitors of the Axin degradation as potent drugs to reduce the concentration of β-catenin (Chen et al., 2009a; Fearon, 2009; Huang et al., 2009; Liu and He, 2010; Zhang et al., 2011). It has been discussed that the low Axin concentration does not only limit the β-catenin degradation but have another implication. As its concentration is very low, the concentrations of other pathway components are only slightly changed by binding Axin, leading to the isolation of the Wnt/β-catenin pathway by avoiding a perturbation of other pathways (Lee et al., 2003).

Further analyses of the core-model have led to the experimentally confirmed prediction that the scaffold proteins APC and Axin contribute differently to the formation of the destruction complex. While APC binds its interaction partners in an ordered manner, Axin binds them randomly. Moreover, it has been discussed that variations in the APC level can be compensated by an APC-dependent Axin degradation. This offers the possibility to stabilise the signalling pathway under conditions with decreased APC levels. The theoretical investigation has also indicated the potential importance of the non-Axin-dependent β-catenin degradation under conditions with low APC concentration (Lee et al., 2003).

Further analyses and insights regarding the core-model

While the sensitivity analysis published by Lee and colleagues (Lee et al., 2003) has been performed with respect to the unstimulated and stimulated steady state, further sensitivity analyses also investigate the impact on other signalling characteristics, such as the β-catenin amplitude or the signalling time. In a sensitivity analysis of a transiently stimulated system, the effects of different perturbation strengths on the steady states as well as on the amplitude, the duration, the signalling time and the integrated response have been calculated (Wawra et al., 2007). The analysis has shown that changes of the protein concentrations have a high impact on the amplitude and the integrated response, whereas the duration and the signalling time are quite robust against perturbations. Additionally, the effects of multiple perturbations on the examined characteristics have been investigated. The analysis has shown that the parameters that have a high impact if perturbed alone also have a high impact under simultaneous variations of the parameters.

A recently published combined experimental and theoretical analysis has focussed on the cellular readout of the Wnt/β-catenin pathway (Goentoro and Kirschner, 2009). Based on a dimensional analysis of the core-model, the reactions have been divided into three subgroups: input, degradation and synthesis. These subgroups differ in part from the submodules introduced by Lee and colleagues (Lee et al., 2003). In Appendix A, Figure A.1, schemata of the core-model are shown, illustrating the modules identified by Lee and colleagues (Figure A.1A) as well as Goentoro and colleague (Figure A.1B). A detailed theoretical investigation by Goentoro and colleague has shown that the fold-change of β-catenin, defined as the ratio of the stimulated to the unstimulated steady state concentration, is quite robust against small perturbations in the degradation module. In contrast, other signalling characteristics such as the absolute level of β-catenin, the difference between stimulated and unstimulated steady state, the response time, the integrated level, the integrated difference and the integrated fold-change are sensitive towards perturbations in all modules. Therefore, the β-catenin fold-change has been considered to be the relevant readout of the signalling pathway. In this way, variations in concentrations or kinetic parameters can be buffered to a certain extent. It has been proven experimentally that the fold-change of β-catenin is relevant for the phenotype. Experiments for two target genes have confirmed that their expression is insensitive to moderate perturbations in the degradation module (Goentoro and Kirschner, 2009).

Further investigations of the system have not only focussed on sensitivity analyses. The dynamics of β-catenin is determined by the binding affinities of the pathway components. Those affinities can be altered by modifications. For example, the binding of Axin and GSK3 can be enhanced by the phosphorylation of Axin by CK1 resulting in a lower TCF-mediated transcription activity. It has been demonstrated experimentally that the protein phosphatase PP1 has an opposite effect according to model results (Luo et al., 2007).

Analyses of transcriptional feedback loops and crosstalk to other signalling pathways
Several gene products of the Wnt/β-catenin pathway participate in the pathway itself, offering the possibility for positive and negative autoregulation (see Figure 2.1). The feedback loop via Axin-2 has been included in several modelling approaches (Cho et al., 2006; Goldbeter and Pourquie, 2008; Jensen et al., 2010; Wawra et al., 2007). Its effect with respect to the β-catenin steady state has been studied in various mutants (Cho et al., 2006). There, the increase of the unstimulated β-catenin steady state arising from mutations has been discussed as a

balanced outcome between pathway activation by the mutation and increased pathway inhibition by the Axin-2 feedback.

Transcriptional feedbacks may cause oscillatory behaviour (Monk, 2003). Interestingly, Axin-2 has been shown to play a critical role in the segmentation clock which is a molecular oscillator determining the timing of the somitogenesis in vertebrates. Beside the Wnt/β-catenin pathway, other signalling pathways such as the Notch and the FGF (fibroblast growth factor) pathways are involved in that clock (Aulehla et al., 2003; Dequeant et al., 2006). Several models focus on the occurrence and characteristics of oscillations in Wnt/β-catenin signalling. The work of Wawra and colleagues extended the core-model of Lee and colleagues by the feedback loops via Axin-2 and Dkk (Wawra et al., 2007). These two feedback loops reinforce their effects on the β-catenin steady state and the oscillations. It has been shown that for sustained oscillations, the involvement of the two feedback loops is not sufficient but also the parameters of the core-model, especially the β-catenin throughput, have to be adapted. The cycle duration of stable oscillations depends on the β-catenin and Axin throughput, the time delay introduced for the transcription, the translation and the splicing and the Hill coefficient that is included in the reactions of gene activation. To obtain stable oscillations with cycle durations that are in accordance with experimental observations (Aulehla and Pourquié, 2008; Aulehla et al., 2003; Dequeant et al., 2006; Dequeant and Pourquie, 2008), further mechanisms are necessary, such as additional feedback loops or interactions with other pathways. In a reduced model of the canonical Wnt/β-catenin pathway, Jensen and co-workers observed oscillations with periods corresponding to the experiments (Jensen et al., 2010).

The possibility for oscillations has also been analysed in models combining the transcriptional feedback via Axin-2 and crosstalks to Notch and FGF signalling. These approaches describe the Wnt/β-catenin pathway with reduced models, using either ODEs (Goldbeter and Pourquie, 2008) or delay differential equations (Rodriguez-Gonzalez et al., 2007). Oscillations can arise in the separated as well as the linked pathways where the crosstalk determines the phase relation (Goldbeter and Pourquie, 2008). Focussing on the crosstalk between Wnt/β-catenin and Notch signalling, it has been shown that target genes of the two pathways, Axin-2 and hes1, influence each other's behaviour. Furthermore, both may act as a master oscillator initiating the oscillation of the whole system (Rodriguez-Gonzalez et al., 2007).

Interactions with other signalling pathways have also been analysed with respect to tumourigenesis. The crosstalk of Wnt/β-catenin and ERK signalling (extracellular-signal-regulated-kinase signalling) has been investigated using the core-model with the modifications introduced by Cho and colleagues (Cho et al., 2006). The analysis of the mathematical model has revealed that the activation of one pathway affects the behaviour of the other signalling system. This has also been shown in experiments: for instance, the activation of the Wnt/β-catenin pathway also leads to a phosphorylation of ERK in both experiments and simulations. The crosstalk of Wnt/β-catenin and ERK signalling results in a positive feedback loop between both pathways and can generate a switch between two stable steady states. As a consequence, mutations in one pathway have an impact on the behaviour of the other pathway. They may lead to a sustained activation of both pathways even in cases in which the stimulus has been removed. This may contribute to tumourigenesis and has to be considered for the development of effective drugs (Kim et al., 2007). The crosstalk and effects of multiple feedbacks within the single pathways have been further investigated with regard to the epithelial-mesenchymal transition (EMT) (Shin et al., 2010). During EMT, cells lose their cell-cell contacts and show a higher mobility. E-cadherin mediates cell adhesion. Therefore, an indicator of epithelial-mesenchymal transition is a reduced E-cadherin expression. The expression is regulated by both the Wnt/β-catenin pathway and ERK signalling. The mathematical model has been analysed with regard to the effects of different combinations of feedback loops working together. The analysis has shown that distinct cellular responses are possible as response of the same stimuli depending on the operating feedback loops. The Wnt/β-catenin pathway and ERK signalling contribute distinctively to the E-cadherin expression. It has been found that RKIP (Raf kinase inhibitor protein), which is a regulator of ERK signalling, plays a key role. This is in accordance with the experimental finding of a RKIP downregulation in tumours (Fu et al., 2003).

For the analysis of larger signalling networks, a further reduction of the Wnt/β-catenin core-model might be advantageous. A minimal model derived by time scale analyses describes the behaviour of β-catenin very well (Mirams et al., 2010).

Models taking the binding of β-catenin and E-cadherin and spatial aspects into account
A model including E-cadherin has been used to address the questions of how the binding of β-catenin and E-cadherin alters the β-catenin availability for the regulation of the gene expression and how the distribution between the two roles of β-catenin is regulated (van Leeuwen et al., 2007). An increase in the synthesis rate of E-cadherin enhances the cell-cell adhesion and may transiently decrease the expression of β-catenin target genes. In the cases of

mutations in β-catenin or APC, the target gene expression as well as the cell-cell adhesion are increased. This model has been embedded in a multiscale approach combining the subcellular, cellular and tissue level of organisation to investigate the dynamics of intestinal tissue renewal. On the cellular level, the contribution of Wnt/β-catenin signalling to the interactions of neighbouring cells and to gene expression and subsequent progression through the cell cycle has been studied (van Leeuwen et al., 2009).

In wild-type cases and with APC mutations, Wnt/β-catenin signalling has been analysed in single intestinal crypts and multiple connected crypts using a partial differential equation (PDE) model (Murray et al., 2010). According to experimental observations, the model has shown that in the wild-type case Wnt/β-catenin signalling, cell motion and proliferation are well balanced, while in a scenario assuming APC mutations, proliferation occurs along the whole crypt axis. For linked crypts, mutational conditions that allow for invasions of neighbouring crypts have been predicted by the theoretical analyses (Murray et al., 2010).

The interaction of β-catenin and E-cadherin and its impact on tumourigenesis has been analysed in a model including the interactions for a cell and a cell layer (Ramis-Conde et al., 2008). In this model, it has been predicted that more β-catenin is bound to E-cadherin in cells in the centre of a tumour, whereas more β-catenin is available for signalling in cells at its margin. The obtained theoretical β-catenin distribution has confirmed experimental observations (Brabletz et al., 2009). Spatial effects have also been investigated within the scope of epidermal appendages (Sick et al., 2006). For this analysis, a reaction-diffusion model concentrating on the interplay of Wnt and Dkk has been used.

2.1.3 Scope of this thesis

We analyse different aspects of Wnt/β-catenin signalling under wild-type conditions and conditions with mutations of pathway components. In particular, these are APC mutations and a mutation of β-catenin. Mutations can be modelled by the variation of rate constants. This is based on the idea that due to mutations, functional regions of a protein are altered. For example, the polarity or charge of key amino acids may change or the secondary or tertiary structure of the protein may be affected. Mutations can lead to truncated forms of the respective protein. This may lead to altered efficiencies of a protein to modify other proteins or to physically interact with them. The mutations may enhance or weaken the protein's abilities. Mutations can also be modelled by extending the model with a mutated species of the examined protein. Thus, if a heterocygote mutation occurs in the cell, the wild-type protein and its mutated counterpart coexist. Both modelling approaches give important insights into the pathway's function and both are used in this thesis to analyse different aspects.

The starting point of the analyses is the core-model developed by Lee and colleagues (Lee et al., 2003). It is also used for comparisons. The core-model is further analysed with respect to inhibitions under wild-type conditions and conditions in which APC is mutated (section 2.2). The effects of single inhibitions as well as double inhibitions are investigated. This allows for predicting effective intervention points. We ask which inhibitions have the strongest effects. Is the effect of an inhibition similar if the β-catenin steady state or the β-catenin fold-change is considered? Moreover, it is of particular interest whether inhibitions that have strong effects in the wild-type model are also able to affect the signalling characteristics in a model in which APC is mutated.

In section 2.3, the model is extended by including a specific β-catenin mutation. The mutation prevents β-catenin from being phosphorylated by the GSK3. Hence, mutated β-catenin is not degraded via the destruction cycle. The impact of the coexistence of the two β-catenin protein forms on the signalling system is analysed theoretically, and predictions are validated by experiments. They are performed in Rolf Gebhardt's group at University Leipzig. The investigations focus on differences in the steady states and dynamics of wild-type and mutated β-catenin. What could be potent strategies to antagonise the consequences of mutated β-catenin?

In section 2.4, the core-model is extended by the inclusion of the Axin-2 feedback loop and the binding of β-catenin and E-cadherin. These extensions reflect important features that occur in cells but not in cell extracts. The effects of these modifications on the pathway's function are investigated primarily in the wild-type system. What are the consequences of the single modifications? Do they influence each other? The focus is on the impact on the unstimulated steady state of β-catenin and the timing of the increase of the concentration upon pathway activation. This project is done in close cooperation with Andreas Hecht's group at University Freiburg.

2.2 Analysis of the effects of inhibitions

Multiple analyses how inhibitors affect distinct signalling pathways have been performed. Effects of single and multiple perturbations have been investigated experimentally, theoretically and in combined approaches of experiment and theory (Fitzgerald et al., 2006; Geva-Zatorsky et al., 2010; Lehar et al., 2008; Lehar et al., 2007; Nelander et al., 2008). This approach provides information on effective intervention points of the signalling pathway. Effects of single and multiple perturbations have been compared and effects of further perturbations can be predicted. Functional relationships between components and the pathway's functioning can be analysed especially by application of multiple drugs.

In this section, we investigate theoretically the effects of single and double perturbations on the steady state concentrations of β-catenin and its fold-change (ratio between steady state concentration with and without stimulation). The analyses focus on the pathway model without mutations and the model assuming a mutation in APC. These models are referred to as models carrying/bearing an APC mutation or model with an APC mutation in the following. APC is often mutated in cancer especially in colon cancer. Due to mutations the ability of APC to mediate the downregulation of β-catenin is reduced (Fearnhead et al., 2001; Gaspar and Fodde, 2004; Kikuchi, 2003; Polakis, 2000).

We ask whether perturbations strongly affecting the β-catenin steady states also have a strong impact on the β-catenin fold-change. Are there reactions influencing the steady states but do not have an impact on the fold-change and vice versa? What are the consequences of the different APC mutations? Does the efficiency of inhibitors change depending on whether they are applied to a wild-type model or a model carrying an APC mutation? What are the effects if two reactions are inhibited at the same time? The core-model of Wnt/β-catenin signalling is used to address these questions.

2.2.1   Model and method

The mathematical model that is used for the analyses is given by equations A.1 to A.37 (Appendix A). Figure 2.4 shows the underlying reaction scheme again (already shown in Figure 2.3). β-Catenin is produced (reaction 12) and degraded via a destruction-complex-independent mechanism (also referred to as alternative degradation, reaction 13). β-Catenin reversibly binds to the destruction complex APC*/Axin*/GSK3 (reaction 8) and is phosphorylated bound to that complex (reaction 9). After the release from the complex (reaction 10), phosphorylated β-catenin (β-catenin*) is degraded (reaction 11). Furthermore, TCF and APC reversibly bind β-catenin (reactions 16 and 17, respectively). In addition, APC reversibly binds Axin (reaction 7). Axin is produced and degraded (reactions 14 and 15, respectively). GSK3 reversibly binds the APC/Axin complex (reaction 6). Within the APC/Axin/GSK3 complex, GSK3 is able to phosphorylate APC and Axin (reaction 4). They are dephosphorylated via reaction 5. A Wnt stimulus induces the activation of inactive Dsh (Dsh$_i$) (reaction 1). Active Dsh (Dsh$_a$) stimulates the dissociation of GSK3 from the complex APC/Axin/GSK3 (reaction 3). Via reaction 2, Dsh$_a$ is deactivated again.

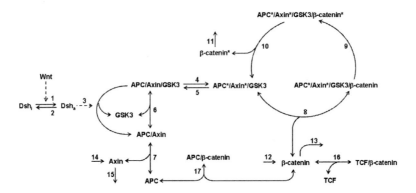

Figure 2.4: Reaction scheme of the core-model of the Wnt/β-catenin pathway.
The model takes the core components into account. See text for details of the single reaction steps. Components in a complex are separated by a slash. The phosphorylation of a component is indicated by an asterisk. Numbers next to arrows denote the number of the particular reaction. One-headed arrows denote reactions taking place in the indicated direction. Dashed arrows represent activation steps and double-headed arrows illustrate binding reactions.

The concentrations of the system's components change due to their production, their degradation, modifications, the binding to other components or their release from a complex. These changes are described by a set of ordinary differential equations (ODEs).
The concentration change of a component i (c_i) is described by:

$$\frac{d(c_i)}{dt} = \sum_{j=1}^{r} n_{ij} \cdot v_j \qquad (2.1)$$

where m is the number of biochemical species with the concentration c_i ($i = 1, ..., m$) and r is the number of reactions taking place in the system with the rates v_j ($j = 1, ..., r$). n_{ij} are the stoichiometric coefficients. The rates are functions of the substrates, reactants and effectors. We assume constant synthesis rates ($v_j = const.$), linear rate equations ($v_j = k_j \cdot c_i$) for dissociations, degradations and modifications such as phosphorylations, dephosphorylations or ubiquitinations. For associations and activation reactions, we use bilinear rate equations ($v_j = k_j \cdot c_i \cdot c_l$). Thereby k_j are the kinetic constants. The model takes 15 variables and 17 processes into account. The equations are provided in Appendix A. The kinetic parameters are provided in Table 5, Appendix A. In the entire chapter 2, the same parameter set is used. Therefore, also dynamical aspects are taken into consideration as they are of interest in the sections 2.3 and 2.4. The basis of the parameter set used in this thesis is the parameter set introduced by Lee and colleagues (Lee et al., 2003) and the modified parameter set published by Cho and colleagues (Cho et al., 2006). The parameters introduced by Lee and colleagues have been measured in *Xenopus* oocyte extracts or have been adjusted to *Xenopus* oocyte extracts data. Fast equilibria for binding reactions have been assumed. In general, Cho and colleagues use the parameter set published by Lee and colleagues but neglect the assumption of fast equilibria. We follow the approach of Cho and colleagues. If no fast binding reactions are assumed, the dynamics of the system slow down. As Cho and colleagues have merely considered the unstimulated β-catenin steady states they could disregard this effect. In section 2.2, we investigate the steady states and fold-change. For these analyses, the slow down of the system would not be of interest. However, in the following sections (sections 2.3 and 2.4), we also analyse the system's dynamics. This is affected by neglecting the fast equilibria. Although we mainly focus on the qualitative properties and not on the exact time points when maxima or steady state concentrations are reached, the system is rescaled to get temporal behaviours of the system that are in a reasonable time frame. To achieve this rescaling, all rate constants are multiplied by a factor of ten.

The total concentrations of APC, Dsh, GSK3 and TCF obey conservation relations and do not change over time. The values are provided in Table 7, Appendix A.

Unbound, unmodified β-catenin is considered to be the pathway readout. We are interested in the effects of parameter changes by 50% on its steady state concentrations. Furthermore, effects on the β-catenin fold-change are analysed which is considered to be the relevant pathway outcome (Goentoro and Kirschner, 2009).

The steady state is calculated by the following formula:

$$\frac{d(c_i^{\text{steady state}})}{dt} = 0 \, .$$

(2.2)

The fold-change is the ratio between stimulated and unstimulated steady state:

$$fold-change = \frac{stimulated\ steady\ state}{unstimulated\ steady\ state} \, .$$

(2.3)

The effects of the changes of parameter j are analysed by calculating the percentaged change of the examined characteristic $S_j^{charact}$ (*charact*: unstimulated steady state, stimulated steady state or fold-change). The change of the characteristic is calculated by:

$$S_j^{charact} = \frac{d(charact)}{charact} \, .$$

(2.4)

The higher the value of $S_j^{charact}$, the stronger the effect of the respective parameter j is. In the following, this is also referred to as the strength of effect. If parameter changes results in $S_j^{charact}$ values that have the same sign, the type of effect mediated by the parameter changes is the same. $S_j^{charact}$ values of unequal signs denote that the mediated types of effect differ.

Besides the consequences of single perturbations, the effects of perturbing two reactions simultaneously are investigated. This is also referred to as double inhibition or double perturbation in the following.

Effects of perturbations are not only interesting under healthy that means wild-type conditions, but also under abnormal conditions. Therefore, we analyse the effects of perturbations in systems with APC mutations. The APC mutations and corresponding parameters are adapted following the publication by Cho and colleagues (Cho et al., 2006). A mutation of APC affects its interactions with Axin or β-catenin. It alters the dissociation of the APC-containing complexes APC/Axin, APC/β-catenin and APC*/Axin*/GSK3/β-catenin. Further details are given in section 2.2.3. The parameters of the individual APC mutations are provided in Table 6, Appendix A.

In this section, the model using the unmodified parameter set is referred to as wild-type model.

2.2.2 Analyses of the wild-type model

Inhibition of individual processes

The effects of the inhibition of each reaction rate on the unstimulated and stimulated steady state concentration of β-catenin (Figure 2.5, row i and row ii, respectively) as well as on the β-catenin fold-change (Figure 2.5, row iii) are investigated.

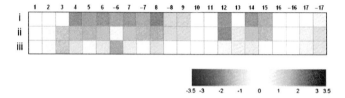

Figure 2.5: Effects of inhibitions on the β-catenin steady states and the fold-change.
The rate constants are decreases by 50%. The effects on the unstimulated (row i) and the stimulated steady state (row ii) as well as on the fold-change (row iii) are shown.
White: no impact; green: an inhibition leads to a lower steady state (fold-change); red: an inhibition leads to a higher steady state (fold-change); the darker the colour, the higher the impact (see legend). The numbers correspond to the numbers of the processes in Figure 2.4. In the case of reversible reactions, the complex formation is denoted with a positive sign; the dissociation of the complex is labelled with the negative reaction number.

Figure 2.5 demonstrates that the impact of the single reactions on the unstimulated and the stimulated β-catenin steady state is very similar (compare rows i and ii). Several reactions have a strong impact on the β-catenin steady states with similar strength, e.g. the phosphorylation of APC and Axin within the complex APC/Axin/GSK3 (reaction 4), the binding of GSK3 to APC/Axin (reaction 6), the binding of APC and Axin (reaction 7), the binding of β-catenin to the destruction complex (reaction 8) and the production of Axin (reaction 14). Other reactions have no impact like the degradation of phosphorylated β-catenin (reaction 11) and the formation and dissociation of the TCF/β-catenin complex (reaction 16). The strength of effect changes slightly due to a stimulus but the type of effect (increasing/decreasing the steady state) stays the same. The exceptions are reactions that do not take place without pathway stimulation. These are the reactions that take Dsh into account, in particular, the activation and deactivation of Dsh (reactions 1 and 2, respectively) and the Dsh-dependent release of GSK3 from the APC/Axin/GSK3 complex (reaction 3).

The analysis of the effects on the β-catenin fold-change (Figure 2.5, row iii) reveals that the type of effect of eleven reactions changes in comparison to the type of effect of the inhibition on the steady states (Figure 2.5, row i and ii). These are the reactions of the formation of the destruction complex and the passing through the destruction cycle (reactions 4 to 9) as well as the synthesis and degradation of Axin (reactions 14 and 15, respectively). These reactions are referred to as destruction reactions in the following.

In comparison to the effects on the unstimulated steady state (row i), the absolute impact on the fold-change (row iii) of eleven reactions decreases. In comparison to the stimulated (row ii) steady state, the absolute impact on the fold-change of 14 reactions decreases. In part, these differences are quite small. The scale in Figure 2.5 covers a large range. Therefore, some differences are hard to recognise. As an example, the values $S_j^{charact}$ of the dissociation of the APC/β-catenin complex (reaction -17) are given: $S_{-17}^{unstimulated\ steady\ state} \approx 0.06$, $S_{-17}^{stimulated\ steady\ state} \approx 0.27$ and $S_{-17}^{fold-change} \approx 0.2$. Hence, although this is not obvious in Figure 2.5, the impact of the dissociation of the APC/β-catenin complex (reaction -17) on the stimulated steady state is larger than on the fold-change.

The reactions whose absolute effects on the unstimulated steady state (row i) are larger than their absolute effects on the fold-change (row iii) are: the production of β-catenin (reaction 12) as well as the destruction reactions (reactions 4 to 9, 14 and 15) with exception of the release of GSK3 from APC/Axin/GSK3 (reaction -6). The impact of these eleven reactions on the stimulated steady state is also larger than their impact on the fold-change. In addition, the impact of the alternative β-catenin degradation (reaction 13) and the formation and dissociation of the APC/β-catenin complex (reactions 17 and -17, respectively) on the fold-change is lower than their impact on the stimulated steady state. In contrast, the impact of the Dsh-independent GSK3 release from the complex APC/Axin/GSK3 (reaction -6) on the fold-change is much stronger than that on the β-catenin steady state. Overall, the strongest impact on the fold-change is mediated by the Dsh-dependent and -independent dissociation of GSK3 from APC/Axin/GSK3 (reactions 3 and -6, respectively). This is due to the fact that the Dsh-dependent release only affects the stimulated steady state while the Dsh-independent dissociation has a strong impact on the unstimulated but only a small impact on the stimulated β-catenin steady state. Therefore, an inhibition of the respective processes affects either the unstimulated or the stimulated steady state but not both steady state levels. This leads to a strong impact on the fold-change.

In summary, the analysis reveals that independent of whether the unstimulated or the stimulated steady states is considered, the type of effect of the individual processes remains the same. However, the strength of effect can differ. If the effects on the fold-change are compared to the effects on the steady states, they also differ in the strength in various cases. In addition, differences in the kind of effect can be observed.

In the following analyses, only the unstimulated steady state and the fold-change of β-catenin are considered.

Effects of double inhibitions in the wild-type model

We analyse the effects of double inhibitions that means we investigate the impact on the unstimulated steady state (Figure 2.6A) and fold-change (Figure 2.6B) if two reactions are inhibited at the same time.

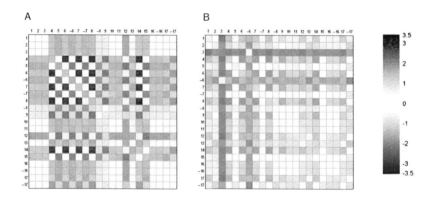

Figure 2.6: Effects of double inhibitions in the wild-type model.

The effects of 50% inhibition of two reactions on (A) the unstimulated steady state and (B) the β-catenin fold-change are shown.

White: no impact; green: an inhibition leads to a lower steady state (fold-change); red: an inhibition leads to a higher steady state (fold-change); the darker the colour, the higher the impact (see legend).

For all non-diagonal elements, the reactions given for the corresponding row and column are inhibited. On the diagonal, reactions are inhibited once. Thus, the diagonal is consistent with the effects of the single inhibition (Figure 2.5, row i or Figure 2.5, row ii). The numbers correspond to the numbers of the processes in Figure 2.4. In the case of reversible reactions, the complex formation is denoted with a positive sign; the dissociation of the complex is labelled with the negative reaction number.

The analysis presented in Figure 2.6A demonstrates that the effects of two combined perturbations can increase or decrease their individual effects on the unstimulated β-catenin steady state. No remarkable synergistic effects occur. Therefore, they are not quantified or analysed in particular.

Inhibiting two destruction reactions simultaneously leads to a stronger effect if both individual inhibitions have the same type of effect (increase or decrease). This occurs e.g. if binding of GSK3 and APC/Axin (reaction 6) and the phosphorylation of APC and Axin within the complex APC/Axin/GSK3 (reaction 4) are inhibited in combination. If two individual inhibitions have effects of the opposite type and of different strength, the inhibition of both reactions at the same time leads to an effect that is weaker than the stronger effect of the individual inhibitions. However, the stronger type of effect of the individual inhibitions remains the type of effect of the combined inhibition. One example is the simultaneous inhibition of the phosphorylation of APC and Axin within the complex APC/Axin/GSK3 (reaction 4) and the binding of APC and β-catenin (reaction 17). If both inhibitions have effects of the same strength but opposite types, the effects annihilate each other. This occurs for instance if the phosphorylation and dephosphorylation of APC and Axin within the complex APC/Axin/GSK3 (reactions 4 and 5, respectively) are inhibited at the same time.

Regarding the fold-change (Figure 2.6B), especially the two reactions describing the Dsh-dependent and -independent release of GSK3 from APC/Axin/GSK3 (reactions 3 and –6, respectively) have strong impacts. Similar to the analysis with respect to the unstimulated steady state of β-catenin, effects of two inhibitions can enhance, compensate or diminish each other's effect on the β-catenin fold-change.

As shown in the case of single inhibition (Figure 2.5), the types of effect mediated by a double inhibition can be the same with respect to the steady state and the fold-change. An example is the inhibition of the phosphorylation of APC and Axin bound to GSK3 (reaction 4) and the alternative β-catenin degradation (reaction 13) at the same time. This causes a higher steady state as well as a higher β-catenin fold-change. On the contrary, inhibiting two reactions simultaneously can also cause opposite types of effects on the unstimulated β-catenin steady state and the β-catenin fold-change. If the alternative β-catenin degradation (reaction 13) is inhibited in combination with the dephosphorylation of APC and Axin bound to GSK3 (reaction 5), it results in a lower steady state but in a higher fold-change.

2.2.3 Effects of inhibitions of processes in the presence of APC mutations

In various diseases, the Wnt/β-catenin signalling is found to be dysregulated due to mutations of pathway components that lead to a perturbation of the pathway. Pharmaceutical drugs are under investigation that act as pathway regulators and that can be used to treat these diseases. It has been and still is the focus of many research groups to test small molecules that influence the pathway to find new potent regulators (see e.g. list of small molecules in Wnt signalling on the Wnt homepage wnt.stanford.edu) (Barker and Clevers, 2006; Chen et al., 2010a; Dodge and Lum, 2010). Ideally, these drugs should affect the system in which a mutation occurs but have little or no effects on the wild-type system. Hence, an analysis of the effects of inhibitions under mutated conditions is of particular interest.

APC is a protein that is often mutated in different kinds of cancer such as colorectal cancer or hepatocellular carcinoma (Fearnhead et al., 2001; Gaspar and Fodde, 2004; Kikuchi, 2003; Polakis, 2000). It is a large protein of 311 kDa, consisting of 2843 amino acids. APC contains Serine-Alanine-Methionine-Proline (SAMP) repeats which are binding sites for Axin, 20 amino acid repeats (20 aa) and 15 amino acid repeats (15 aa) that are essential for the β-catenin binding and degradation, and a recently discovered β-catenin inhibitory domain (CID) (Behrens et al., 1998; Eklof Spink et al., 2000; Kohler et al., 2010). A schematic representation of the APC protein is shown in Figure 2.7.

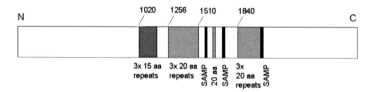

Figure 2.7: Scheme of the APC protein.
The schematic representation of APC illustrates the domains involved in binding Axin and β-catenin. SAMP repeats are coloured in dark blue, 15 aa repeats are coloured in green, and 20 aa repeats are coloured in light blue. The numbers of single amino acids are given for orientation. N and C mark the N-terminus and C-terminus, respectively. The CID is not shown as the scheme concentrates on the domains considered in the modelling approach.

The effects of various APC mutations on the steady state of β-catenin have been analysed theoretically by Cho and colleagues (Cho et al., 2006). They have published parameter sets of various mutations, mostly APC mutations. In general, Cho and colleagues have based their realisation of the APC mutations on a couple of assumptions which are shortly introduced in the following. i) APC truncations are investigated exclusively. There, the mutation results in the deletion of the C-terminal part of the APC protein. The size of the remaining truncated APC protein depends on the location of the mutation. ii) APC mutations have an effect on the dissociation constant K of APC and its binding partners Axin and β-catenin. The dissociation constant is calculated by $K = k_-/k_+$ where k_- is the dissociation rate and k_+ is the association rate. Cho and colleagues have considered that the mutation in APC exclusively affects the dissociation rate and have no effect on the association rate. They have argued that the association rate is mainly determined by diffusion and therefore not affected by the mutation. iii) It is assumed that the loss of one binding domain contributes to an increase of the particular dissociation rate. Cho and colleagues have argued that each binding domain affects the change of the free energy ΔG by an equal amount. It holds that ΔG is proportional to $\ln K$. Hence, the loss of one binding domain caused a multiplicative increase of the respective dissociation rate. Due to a lack of detailed information, allosteric effects are not considered.

For the representation of the distinct APC mutations in the mathematical model, the parameter values of the dissociation of the APC/Axin (reaction −7), the APC/β-catenin (reaction −17) and the APC*/Axin*/GSK3/β-catenin complexes (reaction −8) are altered. If a SAMP repeat is removed, the parameter values of the dissociation of APC and Axin (reaction −7) is increased. If a 20 aa repeat is removed, the rate constant of the release of unmodified β-catenin from the destruction complex (reaction −8) is changed and in the case of the truncation of a 15 aa repeat, the parameter value of the dissociation of the APC/β-catenin complex (reaction −17) is changed.

For this inhibition analysis, four APC mutations are selected. According to the nomenclature of Cho and colleagues, these are the APC mutations m1, m5, m10 and m13. The respective number of the mutation corresponds to the number of domains that are removed. Assuming that it holds that the more domains are removed, the stronger the APC mutation is, the mutations are also referred to as weak (m1), intermediate (m5), strong (m10) and strongest (m13) mutation. The parameters of the individual mutations are provided in Table 6, Appendix A. The models with the APC mutations are referred to as m1 model, m5 model, m10 model and m13 model depending on the occurring APC mutation.

We ask whether a single or double inhibition of pathway processes has the same consequences on the β-catenin steady state or fold-change in a model bearing an APC mutation and in the wild-type model. Moreover we want to know whether the strength of the mutation influences the effect of the inhibition.

Effects of APC mutations on the unstimulated steady state and the fold-change of β-catenin

Due to the mutation in APC, the dissociation of the APC-containing complexes APC/Axin, APC/β-catenin and APC*/Axin*/GSK3/β-catenin is enhanced. As a result, less β-catenin is degraded via the destruction cycle. In Figure 2.8, the (A) unstimulated β-catenin steady states and the (B) fold-changes are presented calculated in the wild-type model and the models with the APC mutations m1, m5, m10 or m13.

A   B

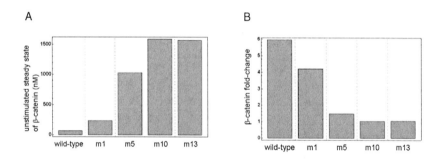

Figure 2.8: Unstimulated β-catenin steady state and fold-change in the wild-type model and model carrying APC mutations

In (A) the unstimulated steady state concentration of β-catenin and in (B) the β-catenin fold-change is shown. The values are calculated for the wild-type model and systems bearing the APC mutations m1, m5, m10 or m13. The bars are labelled with the abbreviation of the particular mutation.

Figure 2.8 reveals that the unstimulated steady state and the fold-change are affected if APC is mutated. Compared to the wild-type model, the steady state is increased and the fold-change is decreased in the systems carrying APC mutations. From mutation m1 to m10 it holds that the stronger the APC mutation, the higher the steady state concentration and the lower the fold-change. The calculations of the m13 model show a deviation of this general trend. Compared to the m10 model, the steady state concentration does not increase but slightly decreases. The fold-change in the m13 model is marginally increased compared to the m10 system, but almost as low as in this model.

In the mutants m1, m5 and m10, the dissociation of APC/Axin (reaction −7) and the dissociation of β-catenin from the destruction complex (reaction −8) are affected. The more interaction domains are deleted, the more the APC/Axin and APC*/Axin*/GSK3/β-catenin complexes dissociate. In consequence, less destruction complex is built and less β-catenin is degraded via the destruction cycle. The β-catenin steady state concentration increases (Figure 2.8A). If a stimulus is applied, the β-catenin concentration increases. A fold-change is induced. Since the unstimulated steady state concentration of β-catenin is already increased due to the APC mutation, the stimulus cannot induce a fold-change as strong as in the wild-type system. The fold-change decreases with increasing strength of the APC mutation (Figure 2.8B).

The calculations in the m13 system reveal an additional characteristic. Compared to the wild-type, the m1 and the m5 model, the APC mutation m13 causes a higher β-catenin steady state and a lower fold-change. However, in comparison to the m10 system, the steady state is slightly decreased from ~1585 nM to ~1567 nM while the fold-change is marginally increased from ~1.03 in the m10 model to ~1.04 in the m13 model. In mutant m10, all interaction domains related to the dissociation of the APC/Axin complex (reaction −7) and the release of β-catenin from the destruction complex (reaction −8) are removed. Mutation m13 is realised by eliminating all interaction domains, including the domains mediating the dissociation of the APC/β-catenin complex (reaction −17). For very high β-catenin concentrations, the formation of the APC/β-catenin complex acts as a positive feedback on β-catenin (Goentoro and Kirschner, 2009). β-Catenin sequesters APC within the APC/β-catenin complex, leading to a decreased availability of APC for the formation of the destruction complex. Consequently, less β-catenin is degraded via the destruction cycle. In the APC mutation m13, the domains associated with the dissociation of the APC/β-catenin complex are removed. Hence, the positive feedback is diminished, leading to a slightly decreased unstimulated β-catenin steady state and a slightly increased fold-change compared to the m10 system. Thus, although APC is stronger truncated in the m13 than in the m10 model, the effects on the unstimulated steady state and the fold-change of β-catenin are smaller in the m13 than in m10 model.

From this analyses we conclude that compared to the wild-type model the unstimulated β-catenin steady state is increased in a model carrying an APC mutation. In addition, in the presence of an APC mutation, the pathway stimulation does not induce a fold-change as strong as under wild-type conditions.

Inhibition of individual processes in the presence of an APC mutation
In Figure 2.9, the effects on the unstimulated steady state are shown for the wild-type model (row i) and the model carrying APC mutations of different strength (row ii to v).

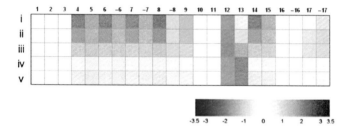

Figure 2.9: Effects of inhibitions on the unstimulated β-catenin steady state in the absence and presence of APC mutations.
The effects of 50% inhibition of an individual reaction are calculated for the wild-type model (row i), the m1 model (row ii), the m5 model (row iii), the m10 model (row iv), and the m13 model (row v). Parameters of the particular mutation are adapted from (Cho et al., 2006). They are given in Table 6, Appendix A.
White: no impact; green: an inhibition leads to a lower steady state; red: an inhibition leads to a higher steady state; the darker the colour, the higher the impact (see legend).
The numbers correspond to the numbers of the processes in Figure 2.4. In the case of reversible reactions, the complex formation is denoted with a positive sign; the dissociation of the complex is labelled with the negative reaction number.

The comparison of the effects of single inhibitions on the unstimulated β-catenin steady state under wild-type and mutated conditions reveals that the impact of the reactions in the wild-type model (Figure 2.9, row i) and the m1 model (Figure 2.9, row ii) is very similar. The changes become more apparent with stronger APC mutations (rows iii to v in Figure 2.9). In general it holds for all APC mutations that the impact of the destruction reactions (reactions 4 to 9, 14 and 15) decreases compared to their impact in the wild-type model. At the same time, the impact of the alternative β-catenin degradation (reaction 13) increases considerably. For the m1 (row i) to m10 system (row iv) the impact increases with increasing strength of the APC mutation. In the m10 system (row iv), the perturbation of the synthesis rate or the alternative degradation rate of β-catenin (reactions 12 and 13, respectively) affects the unstimulated steady state concentration of β-catenin. Inhibiting any other reaction has almost no effect on this steady state level. This is due to the fact that less destruction complex is formed and the binding of β-catenin to the destruction complex is less efficient if APC is mutated. Hence, a lower amount of β-catenin is degraded via the destruction cycle, and the inhibition of a destruction reaction has a smaller effect compared to its effect in the wild-type model (row i). Since less β-catenin is degraded via the destruction cycle, more β-catenin is degraded by the alternative degradation (reaction 13). Its impact increases. The lower the impact of the destruction reactions, the stronger the impact of the alternative β-catenin degradation.

There is one exception of the rule that the stronger the mutation (m1 to m10), the lower the impact of the degradation reactions. In the m5 model (row iii), the impact of the inhibition of the β-catenin release from the destruction complex (reaction –8) is slightly increased in comparison to the wild-type and the other models with APC mutations. This is caused by the increase of the rate constant of this reaction for the realisation of the m5 mutation.

In the m13 model, the impact of the destruction reactions is decreased compared to the wild-type model (compare rows i and v) but slightly increased in comparison to the impact in the m10 system (compare rows iv and v). In the m13 model, the dissociation of the APC/β-catenin complex (reaction 17) is enhanced. Therefore, the effect of the positive feedback of APC/β-catenin is attenuated. Less APC is bound in the APC/β-catenin complex. Compared to the m10 model, more APC is available to participate in the destruction complex. In consequence, the destruction reactions has a slightly stronger effect in the m13 system than in the m10 system. Hence, although more domains are deleted in the APC mutant m13 than in mutant m10, the impact mediated by the inhibition of the destruction reactions in the m13 model is higher than in the m10 model.

Compared to the impact in the wild-type model (row i), the impact of the formation and dissociation of the APC/β-catenin complex (reactions 17 and –17, respectively) slightly increases in the m1 and m5 model (rows ii and iii, respectively) but decreases under the conditions of the stronger APC mutations m10 and m13 (rows iv and v, respectively). In the wild-type model, the impact of these reactions is quite small. Due to an APC mutation, the concentration of β-catenin increases and the impact of the formation and dissociation of the

APC/β-catenin complex increases. The higher the β-catenin concentration, the higher the effect of the positive feedback is. If the mutation is weak, the inhibition of the formation of the APC/β-catenin complex increases the concentration of APC that is available for the formation of the destruction complex. The impact of the inhibition of the binding of APC and β-catenin increases. If the mutation is stronger, the effect resulting from the inhibition of the formation of the APC/β-catenin complex is weaker. The system is not able to utilise the higher amount of APC available for the destruction complex as less destruction complex is formed due to the mutation.

The impact of the synthesis of β-catenin (reaction 12) is almost unchanged in the different models. Hence, the absolute impact of the β-catenin production is almost the same independent of whether an APC mutation occurs or not. Its relative proportion of the overall impact increases as the impact of the majority of processes decreases, for instance the impact of the destruction reactions (reactions 4 to 9, 14 and 15). Above in this paragraph it is already argued that in the case of strong APC mutations, the main effect is mediated by the synthesis (reaction 12) and alternative degradation (reaction 13) of β-catenin.

The degradation of phosphorylated β-catenin (reaction 11) as well as the formation and dissociation of the TCF/β-catenin complex (reactions 16 and –16, respectively) do not have an impact on the unstimulated steady state concentration of β-catenin. This holds for the wild-type model and the models carrying APC mutations.

Taken together the analysis shows that while the inhibition of the alternative β-catenin degradation (reaction 13) has only a little effect on the unstimulated β-catenin steady state in the wild-type model, such a perturbation has a strong effect if APC is mutated. In contrast, the inhibition of the destruction reactions (reactions 4 to 9, 14 and 15) affects the unstimulated β-catenin steady state strongly in the wild-type model but has weaker effects under conditions of APC mutations.

In a further analysis, the effects on the fold-change in the wild-type model and the models with APC mutations are examined. This is shown in Figure 2.10.

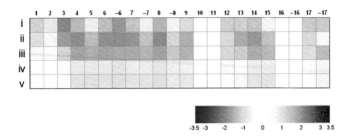

Figure 2.10: Effects of perturbations on the β-catenin fold-change in the absence and presence of APC mutations.

The effects of 50% inhibition of a single reaction are calculated for the wild-type model (row i), the m1 model (row ii), the m5 model (row iii), the m10 model (row iv), and the m13 model (row v). Parameters of the particular mutation are provided in Table 6, Appendix A. The numbers correspond to the numbers of the processes in Figure 2.4. In the case of reversible reactions, the complex formation is denoted with a positive sign; the dissociation of the complex is labelled with the negative reaction number.

White: no impact; green: an inhibition leads to a lower fold-change; red: an inhibition leads to a higher fold-change; the darker the colour, the higher the impact (see legend).

Figure 2.10 shows that in the majority of cases, the effects of single inhibitions on the β-catenin fold-change are of the same type independent of whether APC is mutated or not (compare row i and rows ii to v). Exceptions of this general observation are the types of effect of i) the β-catenin synthesis (reaction 12), and ii) the formation and dissociation of the APC/β-catenin complex (reactions 17 and −17). If these reactions are inhibited, the type of effect is different analysing different APC mutations. The inhibition of the dissociation of the APC/β-catenin complex (reaction −17) in the wild-type model has the opposite effect of the inhibition of this reaction in the models carrying APC mutations. For all analysed APC mutations, the type of effect is the same. In the wild-type model, the inhibition of the dissociation of the complex APC/β-catenin (reaction −17) prevents the release of APC from the APC/β-catenin complex and its participation in the destruction complex. Hence, if the dissociation of the complex APC/β-catenin (reaction −17) is inhibited, the unstimulated and the stimulated steady state levels are higher than under unperturbed conditions. Since the effect on the stimulated steady state concentration of β-catenin is stronger than on the unstimulated steady state level, the fold-change increases. In contrast, in the models in which APC is mutated, this inhibition increases the effect of the respective APC mutation. Less APC is available for the β-catenin degradation via the destruction cycle. As shown in Figure 2.8 it holds in general that the stronger the APC mutation is, the lower the β-catenin fold-change. Thus, if APC is mutated, the inhibition of the dissociation of the complex APC/β-catenin (reaction −17) causes a lower β-catenin fold-change. With one exception, the opposite line of argument holds concerning the inhibition of the formation of the APC/β-catenin complex (reaction 17). The exception is that the type of effect is the same in the wild-type and the m1 model. The change of the type of effect appears for the m5, m10 and m13 models. Like the inhibition of the binding of β-catenin and APC (reaction 17), the inhibition of the β-catenin synthesis (reaction 12) leads to a lower fold-change in the wild-type and m1 models but to an increased β-catenin fold-change in the m5, m10 and m13 systems. These changes of the type of effect is explained by the fact that the APC mutation leads to changes in the unstimulated steady state level of β-catenin (see Figure 2.8). As a consequence, the pathway activation can have different effects. For instance, in the wild-type model, the inhibition of the β-catenin synthesis (reaction 12) causes a smaller fold-change as less β-catenin is produced. In contrast in the m5 system, the inhibition of the β-catenin synthesis leads to an increased fold-change. Due to the mutation, the unstimulated β-catenin steady state is increased. The stimulus is not able to induce a fold-change as strong as in the wild-type model (see Figure 2.8). If the β-catenin synthesis is inhibited in the presence of the m5 mutation, the β-catenin concentration decreases and a stimulus is able to have a stronger effect. The fold-change increases.

Besides changes of the type of effect, the strength of effect differs depending on the particular model. Regarding the β-catenin steady state in the m1, m5 and m10 systems (see Figure 2.9), the conclusion is drawn that the stronger the mutation, the lower the impact of the degradation reactions and the stronger the impact of the alternative β-catenin degradation is. Such a general conclusion cannot be drawn with respect to the β-catenin fold-change. The strength of effect mediated by an inhibition depends on the strength of the considered APC mutation. Comparing the effects mediate by inhibitions in the wild-type (row i) and the m1 model (row ii) reveals that the absolute impact of five reactions decreases, the absolute impact of 13 reactions increases and that the impact of one reaction changes the sign (dissociation of the APC/β-catenin complex, reaction −17). The reactions whose strength of effect decrease are: i) the activation and deactivation of Dsh (reactions 1 and 2, respectively), ii) the Dsh-dependent release of GSK3 from APC/Axin/GSK3 (reaction 3), iii) the β-catenin synthesis (reaction 12), and iv) the formation of the APC/β-catenin complex (reaction 17). Regarding the β-catenin production (reaction 12) and the formation of the APC/β-catenin complex (reaction 17) the same line of argument holds that explains the change of effect of these reactions. In the case of the weak APC mutation, the effect is only diminished leading to a weak change of the strength of effect but not to a change of the type of effect. The reactions involving Dsh (reactions 1 to 3) reflect the strength of the given stimulus. Under wild-type conditions, the stimulus induces a high fold-change. The stimulus has a strong effect, and thus, the reactions reflecting the stimulus have strong effects as well. The stronger the APC mutation is, the weaker the stimulus-induced fold-change (Figure 2.8). In consequence, the impact of the reactions involving Dsh (reactions 1 to 3) decreases.

The absolute strength of effect of the destruction reactions (reaction 4 to 9, 14 and 15) increase if one compares the wild-type (row i) with the m1 and m5 models (rows ii and iii, respectively). The only exception is the Dsh-independent release of GSK3 from APC/Axin/GSK3 (reaction −6) whose strength of effect decreases in the m5 model. The comparison of the m1 and m5 models (rows ii and iii, respectively) with the m10 and m13 models (rows iv and v, respectively) shows that the strength of effect of these reactions decreases. However, in part their impact is increased in comparison to the effects in the wild-type model (row i), for example the impact of the Axin degradation (reaction 15) or the dephosphorylation of APC and Axin bound to GSK3 (reaction 5). The strength of effect of the destruction reactions (reaction 4 to 9, 14 and 15) is higher in the case of a weak and intermediate APC mutation (rows ii and iii, respectively) than under conditions of the strong and strongest APC mutation (rows iv and v, respectively) since in the m1 and m5 models a high amount of β-catenin is degraded via the destruction cycle. The destruction reactions have a strong effect on the β-catenin steady state (Figure 2.9) and on the β-catenin fold-change. In contrast, in the m10 and m13 model, only a small part of β-catenin is degraded via the destruction cycle. The strength of effect of the destruction reactions decreases.

The degradation of phosphorylated β-catenin (reaction 11) and the formation and dissociation of the TCF/β-catenin complex (reactions 16 and −16, respectively) do not have an influence in both cases.

Like the analysis on the unstimulated steady state of β-catenin (Figure 2.9), the analysis in Figure 2.10 shows that the strength of effect on the β-catenin fold-change differs depending on the strength of the APC mutation, but a general conclusion as drawn with respect to the steady state is not possible. The strength of effect of the reactions involving Dsh (reactions 1 to 3) decreases with increasing strength of the APC mutation. The strength of effect of the destruction reactions (reactions 4 to 9, 14 and 15) is higher in the m1 and m5 model than in the m10 and m13 model. For some processes, the strength of effect decreases from the m1 to the m5 model and from the m5 to the m10 model. However, there are other processes whose strength of effect increase from the m1 to the m5 model but decreases if the m10 model is considered. Moreover, the analysis reveals a further feature concerning the β-catenin fold-change. The inhibition of the β-catenin production (reaction 12) and the formation and dissociation of the APC/β-catenin complex (reactions 17 and –17, respectively) can result in different types of effects depending on the particular APC mutation which is investigated.

Effects of double inhibitions in the m5 model

We are also interested in the effects of double inhibitions in the presence of an APC mutation. As example for this analysis, the model of the intermediate APC mutation (m5 model) is considered.

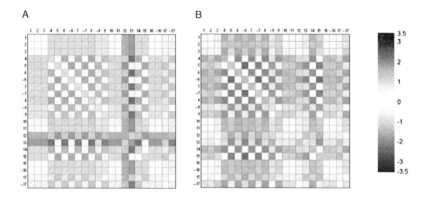

Figure 2.11: Effect of double inhibitions on the unstimulated steady state and fold-change of β-catenin in the m5 model.

The effects on (A) the unstimulated β-catenin steady state and (B) the β-catenin fold-change in the m5 system are shown if two reactions are inhibited at the same time. Parameters of mutation m5 are provided in Table 6, Appendix A. The numbers correspond to the numbers of the processes in Figure 2.4.

For all non-diagonal elements, the reactions given for the corresponding row and column are inhibited. On the diagonal, reactions are inhibited once. Thus, the diagonal is consistent with the effects of the single inhibition (Figure 2.9, row iii or Figure 2.10, row iii).

White: no impact; green: an inhibition leads to a lower fold-change; red: an inhibition leads to a higher fold-change; the darker the colour, the higher the impact (see legend).

The effects of inhibitions of all parameter combinations on the unstimulated β-catenin steady state (Figure 2.11A) as well as on the β-catenin fold-change (Figure 2.11B) are shown. A comparison of the analysis of the m5 model (shown in Figure 2.11A) and the analysis of the wild-type model (Figure 2.6A) reveals that inhibitions have the same type of effect on the steady state independent of whether APC is mutated or not. In comparison to the wild-type model, in the m5 model the strength of impact of the degradation reactions (reactions 4 to 9, 14 and 15) decreases if two reactions are inhibited at the same time. This is also observed in the analysis of the effects of the single inhibitions (Figure 2.9). Hence, the inhibition of two degradation reactions (reactions 4 to 9, 14 and 15) at the same time has a smaller effect in the m5 model than in the wild-type model. The analysis of the single inhibitions (Figure 2.9) reveals that the impact of the alternative β-catenin degradation (reaction 13) increases if APC is mutated. The impact of the β-catenin synthesis (reaction 12) is very similar in the model without and with an APC mutation. In general, it holds that the inhibition of one of these two reactions in combination with another reaction has the strongest impact on the unstimulated steady state of β-catenin in the m5 model. As stated above (Figure 2.9), the absolute impact of the degradation reactions (reactions 4 to 9, 14 and 15) decreases if APC is mutated. They contribute to a lower amount to the overall impact in the case of a double inhibition. If the β-catenin production (reaction 12) is inhibited in combination with a degradation reaction whose effect is of the same type (reactions 5, −6, −7, −8, 15), the total effect of the double inhibition is slightly smaller in comparison to the effect of the double inhibition in the wild-type model (Figure 2.6). In contrast, if the β-catenin synthesis (reaction 12) is inhibited together with a destruction reaction that mediates the opposite kind of effect (reactions 4, 6, 7, 8, 9, 14), the total effect of the double inhibition is stronger than in the wild-type system. However, their total strength of effect is still lower than in the case of inhibiting two reactions that mediate the same kind of effect.

Regarding the β-catenin fold-change, the comparison of the wild-type (Figure 2.6B) and the m5 model (Figure 2.11B) shows that the overall picture of the double inhibition study changes if APC is mutated. The type and strength of the effects resulting from the inhibition of two reactions at the same time change for several combinations of inhibitions. Differences of the type of effect in the different models (with and without APC mutation) are also observed in the single perturbation analysis (Figure 2.10).

In the wild-type system, the inhibition of the Dsh-dependent release of GSK3 from APC/Axin/GSK3 (reaction 3) results in a lower β-catenin fold-change independent of the reaction that is inhibited additionally (Figure 2.6). The only exception of this general observation is the combined inhibition of the Dsh-dependent (reaction 3) and Dsh-independent (reaction –6) release of GSK3 from the complex APC/Axin/GSK3. In the m5 system (Figure 2.11), the effect of the inhibition of the Dsh-dependent dissociation of APC/Axin/GSK3 complex (reaction 3) is weaker than its effect in the wild-type system (Figure 2.6B) and does not dominate the effects of all other inhibitions. Therefore, a higher β-catenin fold-change is achieved if the Dsh-dependent release of GSK3 from APC/Axin/GSK3 is inhibited in combination with reactions whose single inhibitions cause higher fold-changes. These reactions are: i) the synthesis of β-catenin (reaction 12), ii) the alternative degradation of β-catenin (reaction 13), iii) the formation of the APC/β-catenin complex (reaction 17) as well as iv) the processes facilitating the decomposition of the destruction complex. In particular these are i) the dephosphorylation of APC and Axin bound to GSK3 (reaction 5), ii) the Dsh-independent release of GSK3 from APC/Axin/GSK3 complex (reaction –6), iii) the dissociation of APC/Axin (reaction –7), iv) the release of unmodified β-catenin from the destruction complex (reaction –8), and v) the degradation of Axin (reaction 15). Thus, while the double inhibitions results in the same type of effect on the unstimulated β-catenin steady state independent of whether APC is mutated or not, double inhibitions can mediate distinct effects on the β-catenin fold-change in the presence and absence of the intermediate APC mutation m5. While the impact of the destruction reactions (reactions 4 to 9, 14 and 15) decreases with respect to the steady state, their impact increases if the fold-change is considered. This is also seen for the analysis of the single inhibition (Figure 2.10).

Various combinations of inhibitions increase the β-catenin fold-change stronger in the m5 than in the wild-type model (compare Figure 2.6B and Figure 2.11B). However in general, the fold-change in the m5 system remains on a lower level than the fold-change in the wild-type model. The wild-type fold-change is ~5.9 and the fold-change in the unperturbed m5 model is ~1.48 (Figure 2.8). By a double inhibition of 50%, the fold-change in the m5 model can be increased the most to ~3.7. This is a markable increase compared to the fold-change in the unperturbed m5 model, but the β-catenin fold-change is still not as high as under wild-type conditions. The unstimulated β-catenin steady state in the m5 model is always higher than in the wild-type model. The β-catenin steady state is ~68.7 nM in the wild-type model. In the unperturbed m5 model, this level is ~1028 nM (Figure 2.8). A double inhibition in the m5 model can cause a decrease in the unstimulated β-catenin steady state but it does not reach a level lower than ~295 nM. Thus, the β-catenin steady state is still elevated at least by a factor of ~4.3. Hence, even a combination of two inhibitors mediating inhibitions of 50% is not able to recover the unstimulated steady state level of β-catenin or the β-catenin fold-change of the wild-type model.

2.2.4   Discussion

In this section it is investigated how inhibitions of processes of the pathway by 50% affect the steady state level of β-catenin and the fold-change. Besides single inhibitions, double inhibitions are examined. The effects are investigated in the wild-type model. More importantly, the effects of parameter changes are analysed in models carrying APC mutations of different strengths. This enables the comparison of the effects of inhibitions i) between the wild-type and a mutated model, and ii) between models carrying different APC mutations. Making predictions of effective inhibitions under different conditions is possible.

For the wild-type model, sensitivity analyses of the steady state (Lee et al., 2003) and the fold-change (Goentoro and Kirschner, 2009) have been published. Their general findings are reproduced in the present study. Effects of combined inhibitions of two reactions or analyses of parameter changes in the models carrying APC mutations are yet not published. An analysis of multiple parameter perturbations concentrates on the effects of the variation of protein concentrations (Wawra et al., 2007).

Inhibitors of Wnt/β-catenin signalling – a comparison of experimental and theoretical findings

As dysregulated Wnt/β-catenin signalling is considered to be related to various diseases and different types of cancer, the development of pathway antagonists is of high interest and the impact of lots of small components has been and still is investigated in experiments (see e.g. list of small molecules in Wnt signalling on the Wnt homepage wnt.stanford.edu) (Barker and Clevers, 2006; Chen et al., 2010a; Dodge and Lum, 2010). Besides molecules affecting the Axin turnover (Chen et al., 2009a; Huang et al., 2009), especially molecules influencing the functioning of the receptor events (Chen et al., 2009b; Lavergne et al., 2010) and molecules having an impact on the interactions of β-catenin and TCF/LEF (Chen et al., 2009c; Park et al., 2005) have been analysed. Furthermore, molecules regulating the alternative β-catenin degradation have been investigated experimentally (Dimitrova et al., 2010; Gwak et al., 2009;

Park et al., 2006). The core-model does not include receptor events and a detailed description of the interaction of β-catenin and TCF/LEF. As the analysis is based on the core-model, effects of those inhibitors are not investigated. Thus, a comparison between experimental observations and model results is not possible.

The effect of the perturbation of the alternative β-catenin degradation is investigated in the presented analysis and has been seen in experiments (Dimitrova et al., 2010; Park et al., 2006). It has been shown experimentally that an activation of the alternative degradation leads to a decrease of the β-catenin concentration. The findings by Gwak and colleagues (Gwak et al., 2009) are in part represented by the model. In the wild-type cell, the activation of the Siah-1-dependent alternative β-catenin degradation leads to a lower β-catenin level in the experiment as well as in the model. In the case of an APC mutation, Gwak and colleagues have not been able to measure a decrease of the β-catenin concentration by activating the Siah-1-dependent alternative degradation. In contrast, the model predicts a strong impact of the regulation of the alternative degradation on the β-catenin steady state. Its impact is even stronger if APC is mutated. This deviation between experimental and theoretical observation can be explained by the fact that the Siah-1-dependent alternative β-catenin degradation is also dependent on APC (Liu et al., 2001). By an APC mutation this function may be affected as well. The effect of an APC mutation on the alternative β-catenin degradation is yet not taken into consideration in the model.

Double inhibitions in the wild-type model

Generally, double inhibitions can be discussed in three ways: i) two inhibitory drugs are applied at the same time, ii) an inhibitory drug is added in the case of an inhibitory mutation, and iii) two inhibitory mutations occur in combination. Here, the results are discussed with respect to the effects of administration of two inhibitors at the same time.

The analyses are used to predict the inhibitions that have the strongest or weakest effect on the unstimulated steady state or fold-change of β-catenin, that act jointly and that counteract one another's effect. The combined application could lead to a stronger or weaker effect than a single inhibition cause. In the case of two inhibitions working together in a wild-type cell, this is also of interest as some inhibitors do not act specifically on one reaction but interfere with different components or at different stages of the pathway. For instance the inhibition of the GSK3, which is often used for pathway activation, does not only influence one reaction of the pathway but affects the pathway at different stages. In the model, the inhibition of the

GSK3 is represented by the inhibition of the phosphorylation of APC and Axin bound to GSK3 (reaction 4) and by the inhibition of the GSK3-mediated phosphorylation of β-catenin bound to the destruction complex (reaction 9). The analysis enables to conclude that these two inhibitions cooperate, increasing the unstimulated steady state of β-catenin to a higher level than a single inhibition cause. However, the increase is not as strong as other double inhibitions could induces, such as the combined inhibition of the APC and Axin phosphorylation bound to GSK3 (reaction 4) and the inhibition of the binding of β-catenin to the destruction complex (reaction 8). Thus, the modelling result leads to the suggestion that drugs affecting these reactions activate the Wnt/β-catenin pathway stronger than the inhibition of the GSK3.

Besides the effects on the unstimulated β-catenin steady state, effects on the β-catenin fold-change are investigated. According to single inhibitions, the overall picture with respect to the fold-change differs from the overall picture concerning the effects on the steady state. Other combinations of inhibited reactions have an influence on the fold-change than on the steady state. The type of effect (increase/decrease) on the β-catenin fold-change can differ from the type of effect on the unstimulated steady state level of β-catenin. For example the effect of a GSK3 inhibitor (inhibition of reactions 4 and 9) causes an increased steady state but a decreased fold-change. This is due to the increased unstimulated β-catenin steady state. A stimulus induces only a small increase of the β-catenin concentration. The β-catenin fold-change decreases. There are also cases in which the type of effect is the same for steady state and fold-change. The inhibition of the dephosphorylation of APC and Axin bound to GSK3 (reaction 5) in combination with the inhibition of the binding of APC and Axin (reaction 7) compensate each other's effects on both the steady state and the fold-change. However, this is not a general feature that combinations of inhibition leading to annihilation in the effects on the unstimulated steady state do also neutralise each other's effect on the fold-change. The simultaneous inhibition of the phosphorylation of APC and Axin bound to GSK3 (reaction 4) and the binding of GSK3 to the APC/Axin complex (reaction 6) has no effect on the unstimulated β-catenin steady state. In contrast, the β-catenin fold-change is decreased by this combination of inhibitions.

Therefore, we stress that before a combination of inhibitors is applied it is necessary to specify the pathway's readout of interest.

Effects of APC mutations on the unstimulated β-catenin steady state and the β-catenin fold-change

The effects of single inhibitions and double inhibitions in the presence of APC mutations are investigated. For this analysis, four APC mutations of different strength (Cho et al., 2006) are selected. The unstimulated β-catenin steady state levels and the β-catenin fold-changes of the wild-type model and the four models with mutated APC (m1, m5, m10, and m13 model) are calculated (Figure 2.8). The calculated steady state concentrations differ from the concentrations calculated by Cho and colleagues (Cho et al., 2006) since they have considered a system including an Axin feedback. However, their general observation that the unstimulated β-catenin steady state increases from the wild-type model to the m10 model and decreases slightly from the m10 model to the m13 model holds also for the model without Axin feedback (Figure 2.8). In consequence, one has to reconsider the conclusion drawn by Cho and colleagues that the slight decline of the unstimulated β-catenin steady state level from the m10 model to the m13 model results from the feedback via Axin. This decline also occurs in the m13 model without the Axin feedback loop. In the analysed m13 model, this decrease of the β-catenin steady state results from the activation of the dissociation of the APC/β-catenin complex (reaction −17). Due to this, less APC is bound in that complex and consequently, more APC is available for the formation of the destruction complex. As the β-catenin level is increased in the models carrying an APC mutation, a high amount of the APC/β-catenin complex is built. If the dissociation of the APC/β-catenin complex (reaction −17) is activated, a high amount of APC is released, leading to a decrease of the β-catenin concentration. In a theoretical analysis reflecting an overexpression of β-catenin, the following effect has been observed: if the concentration of β-catenin is very high, it sequesters APC in the APC/β-catenin complex, reducing the concentration of APC available for the formation of the destruction complex. This is a positive feedback on the β-catenin concentration (Goentoro and Kirschner, 2009). The same mechanism acts in the case of the APC mutation. Hence, the reduced unstimulated β-catenin steady state in the m13 model in comparison to the level in the m10 model is caused by the weaker positive feedback in the m13 system.

In the last years, several studies on the function of the single domains of APC have been published (Kohler et al., 2010; Kohler et al., 2009; Kohler et al., 2008; Munemitsu et al., 1995; Yang et al., 2006). In these publications, the concentration of β-catenin, the ability of mutated APC to bind β-catenin or the transcriptional activity of β-catenin has been reported. The transcriptional activity of β-catenin can be considered as the pool of unbound β-catenin that is present in the cell. It increases if mutated APC instead of wild-type APC is available in the cell (Kohler et al., 2010; Kohler et al., 2009; Kohler et al., 2008; Yang et al., 2006). In general, this activity increases the stronger, the stronger the APC mutation is, that means the more domains of APC are removed or mutated. This is in full accordance with the theoretical findings.

There are new experimental findings that are not taken into consideration by the model. Recent experimental studies have revealed further domains of APC to be involved in the regulation of β-catenin and new functions of known domains. The β-catenin inhibitory domain has been discovered (Kohler et al., 2009) that was not considered in the theoretical analysis by Cho and colleagues (Cho et al., 2006). Thus, their effects are not covered by the theoretical analyses. The experimental investigations on the single domains have elucidated their contributions to the β-catenin degradation. i) It has been shown that the first three 20 amino acid repeats take differentially part in the binding and degradation of β-catenin (Kohler et al., 2008). The strongest effect is mediated by the first repeat while the second repeat has almost no β-catenin binding activity (Kohler et al., 2008; Liu et al., 2006). For the realisation of the distinct mutations, Cho and colleagues have assumed that each repeat contributes equally to the change of the free energy of the binding reactions. ii) Moreover, it has been shown in experiments that the 15 amino acid repeats are involved in the β-catenin degradation (Kohler et al., 2010). Cho and colleagues have proceeded on the assumption that the 15 amino acid repeats bind β-catenin but that they are not involved in the β-catenin degradation. These recent experimental findings provide the possibility to refine the theoretical description of the single APC mutations and to extent the analysis of the effects of inhibitions.

Single inhibitions in a system with an APC mutation

The analysis of a single inhibition in a system with an APC mutation shows that the strength of the effect of the inhibition depends on the strength of the APC mutation. Regarding the unstimulated β-catenin steady state it holds that the stronger the APC mutation is, the stronger the effect of the inhibition of the alternative degradation of β-catenin (reaction 13) and the weaker the effect of the inhibition of a destruction reaction (reaction 4 to 9, 14 and 15). As the APC mutations enhance the dissociations of APC-containing complexes (APC/Axin, APC/β-catenin and APC*/Axin*/GSK3/β-catenin), a lower amount of the destruction complex is formed and less β-catenin is degraded via the destruction cycle. A higher amount of β-catenin is degraded via the alternative degradation (reaction 13), increasing its impact. This implies that drugs affecting the destruction reactions are not effective in each case. While they cause a strong effect in the case of a weak APC mutation, they are not effective if APC is strongly mutated. In the case of these strong APC mutations, drugs affecting other processes are necessary, such as a drug activating the alternative β-catenin degradation.

Regarding the unstimulated β-catenin steady state, the strength of effect changes depending on the presence and the strength of the APC mutation. However, the type of effect stays the same. Concerning the fold-change, the type of effect can change depending on the occurring APC mutation. The reactions whose types of effect change are: i) the β-catenin production (reaction 12), and ii) the formation and dissociation of the APC/β-catenin complex (reaction 17 and −17, respectively). For all other reactions, the kind of effect is the same for each APC mutation. This implies that an inhibitor causing an increase of the β-catenin fold-change in a cell bearing one type of APC mutation can result in a decrease of the fold-change in a system carrying another APC mutation. With respect to the development of effective drugs it is important to know whether a specific drug affects the system in the required way.

Double inhibitions in the m5 system

For treating diseases, multicomponent therapies are common. The application of a combination of two (or even more) drugs could delay the appearance of resistance and may contribute to the necessity of the administrations of lower dosages of one drug, alleviating side effects. Furthermore, the effects of two drugs may not even add but may have more than additive effects. The analyses show that the synergetic effects are very low. Therefore, they are not quantified but the qualitative effects are regarded.

The effects on the unstimulated β-catenin steady state and the β-catenin fold-change are analysed. Although several combinations of inhibitions show a high impact on both characteristics, no combination (with 50% perturbation) exists, leading to the recovery of the steady state and fold-change as observed under wild-type conditions. In each case, the steady state remains increased while the fold-change remains decreased. In the case of an APC mutation, the impact of the alternative β-catenin degradation on the steady state increases. Regarding the β-catenin fold-change, more reactions affect the fold-change to a higher extent compared to the wild-type model. While in the wild-type model the reactions mediating the release of GSK3 from the APC/Axin/GSK3 complex (reactions 3 and –6) have the strongest impact, more reactions have strong impacts on the fold-change if APC is mutated. Compared to the wild-type model, several double inhibitions change their type of effect, meaning that such a double inhibition increases (decreases) the fold-change in the wild-type but decreases (increases) this ratio in the m5 model. Regarding the unstimulated β-catenin steady state, the types of effect stay the same in the wild-type and the m5 system.

Of particular interest are drug combinations that have a weak effect in the wild-type model but a strong effect in the m5 model. This implies that a healthy cell is only slightly affected by the drugs whereas the cell carrying the mutation is strongly influenced. Combinations of inhibitions that lower the unstimulated β-catenin steady state and that have a stronger impact in the m5 than in the wild-type model, are: the inhibition of the β-catenin production (reaction 12) combined with the inhibition of i) the APC and Axin phosphorylation bound to GSK (reaction 4), ii) the binding of GSK3 and APC/Axin (reaction 6), iii) the formation of the APC/Axin complex (reaction 7), iv) the binding of β-catenin to the destruction complex (reaction 8), or v) the production of Axin (reaction 14). The effect of a combined inhibition of these reactions on the fold-change is similar in the wild-type and in the m5 model.

The β-catenin fold-change is lower in the m5 than in the wild-type system. Consequently, combinations of inhibitions are of interest that have only little effects in the wild-type model but increase the fold-change in the m5 model. Numerous combinations of inhibitions exist satisfying this requirement, e.g. inhibiting the dephosphorylation of APC and Axin bound to GSK3 (reaction 5) in combination with the inhibition of i) the dissociation of APC/Axin (reaction –7), ii) the release of β-catenin from the destruction complex (reaction –8), or iii) the degradation of Axin (reaction 15).

2.3   Effects of the coexistence of wild-type and mutated β-catenin on the signalling properties of the pathway

For its proteasome-dependent degradation, β-catenin has to be phosphorylated at specific residues in the aminoterminal domain. First, the priming-phosphorylation at Ser[45] mediated by CK1α takes place. Subsequently, GSK3 phosphorylates β-catenin at Thr[41], Ser[37] and Ser[33]. The phosphorylation of these specific residues enables the recognition of phosphorylated β-catenin by the ubiquitin-ligase complex, leading to the β-catenin degradation. If these residues are not phosphorylated, the ubiquitin-ligase complex is not able to recognise the protein and β-catenin is not degraded.

In various human cancers, mutations occurring in the phosphorylation region of β-catenin have been found. In humans, this region is encoded in exon-3. Due to mutations within the phosphorylation region or the deletion of this domain, β-catenin is not phosphorylated and degraded via the proteasome. This results in cytoplasmic and nuclear accumulation of β-catenin (Austinat et al., 2008; de La Coste et al., 1998; Laurent-Puig et al., 2001; Nhieu et al., 1999; Wong et al., 2001; Zucman-Rossi et al., 2007).

In hepatocellular carcinoma (HCC), one of the most common cancers worldwide, mutations in the β-catenin gene occur with a frequency of 20–30%, in hepatoblastomas even up to 65% (de La Coste et al., 1998; Miyoshi et al., 1998; Takigawa and Brown, 2008; Taniguchi et al., 2002). Axin is another protein that is often mutated in HCC (Satoh et al., 2000; Takigawa and Brown, 2008; Taniguchi et al., 2002). In colorectal cancer, β-catenin mutations have been found. Also mutations in APC occur in this type of tumour (Miyoshi et al., 1992; Salahshor and Woodgett, 2005). In the cases of mutated β-catenin, the mutated protein and the wild-type protein coexist in the cell.

In several experimental studies, the expression of target genes has been measured in healthy livers as well as HCCs carrying either a mutation in β-catenin or mutations in other pathway components for example in Axin (Stahl et al., 2005; Zucman-Rossi et al., 2007). Depending on the presence and the kind of mutation, the gene expression profiles differ. The reason for this might be that mutated β-catenin contributes differently to the gene expression than wild-type β-catenin.

In the analysis presented in section 2.3, we are interested in the effects of mutated β-catenin that is not phosphorylated within the destruction complex. For the theoretical investigation, we extended the core-model of the Wnt/β-catenin signalling pathway by the inclusion of the mutated form of β-catenin and its possible interactions (see section 2.3.1). Both wild-type and mutated β-catenin coexist in the model and act as pathway readouts. We address the question whether the presence of the mutated protein affects the signalling properties of the wild-type protein. Do both β-catenin species behave similarly or are there observable differences if the pathway is stimulated?

The model is analysed and the results are compared with experimental observations. Predictions are made and experiments have been performed for their verification (sections 2.3.2 to 2.3.4) in Rolf Gebhardt's group at the University Leipzig. Prospective analyses will concentrate on the effects of mutated β-catenin on the gene expression of specific genes. Therefore, a minimal model of the extended pathway model is introduced that reflects the main dynamics of the detailed model (section 2.3.5).

2.3.1   Model including mutated β-catenin

Here, we are interested in the effects of mutated β-catenin (β-catenin^{mut}) on the steady states and the temporal behaviour of wild-type β-catenin. The mathematical description is based on the core-model (Lee et al., 2003). In the context of the investigation of mutated β-catenin, the core-model is referred to as wild-type model. Processes of the wild-type model are shown in black in Figure 2.12. To consider the mutated protein, additional processes are taken into account, shown in blue in Figure 2.12. The β-catenin protein with a mutation in exon-3 cannot be phosphorylated and is therefore not degraded via the proteasome. Beside this characteristic we assume the mutated β-catenin form to be involved in similar processes as the wild-type protein since there are no indications that the mutated protein differs in its binding characteristics from the wild-type protein. Mutated β-catenin is produced (reaction 18) and reversibly binds APC (reaction 22), TCF (reaction 21) or the destruction complex APC*/Axin*/GSK3 (reaction 20). Mutated β-catenin is only degraded via the alternative degradation (reaction 19). The model including mutated β-catenin is referred to as mutated model.

Overall, the mutated model contains 19 variables and 22 processes. It is described by a system of differential equations consisting of equations A.1, A.2, A.4–A.6, A.8–A.31 (Appendix A) and equations B.1 to B.14 (Appendix B). Again, constant production rates are assumed. Linear rate equations are assumed for phosphorylations, dephosphorylations, degradations, dissociations and deactivations, and bilinear rate equations for associations and activations. There are no experiment-based indications that the binding characteristics of the mutated protein differ from the binding properties of the wild-type protein. Therefore, the same parameters values are used for the processes independent of whether they are related to wild-type or mutated β-catenin. The kinetic parameter regarding mutated β-catenin are provided in Table 8, Appendix B. For the processes related to wild-type β-catenin, the parameters listed in Table 5, Appendix A, are used.

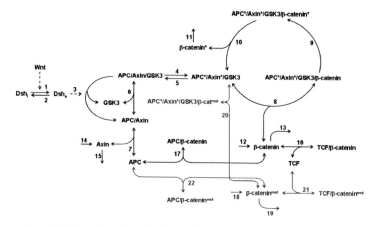

Figure 2.12: Reaction scheme for the Wnt/β-catenin pathway taking into account mutated β-catenin.

The core-model of the Wnt pathway (Lee et al., 2003), processes in black, is extended by including a mutated form of β-catenin (β-catenin^{mut}; β-cat^{mut} in the complex APC*/Axin*/GSK3/β-cat^{mut}). The reactions and components related to mutated β-catenin are shown in blue. Mutated β-catenin is produced and degraded in the destruction-complex-independent manner. It is able to reversibly form complexes with APC, TCF and APC*/Axin*/GSK3.

Components in a complex are separated by a slash. The phosphorylation of a component is indicated by an asterisk. Numbers next to arrows denote the number of the particular reaction. One-headed arrows denote reactions taking place in the indicated direction, dashed arrows represent activation steps and double-headed arrows illustrate binding reactions.

The model enables an analysis of effects if mutated β-catenin and wild-type β-catenin coexist in a system. Experimentally, both protein forms occur together if the cell carries one mutant and one wild-type allele of β-catenin. Another possibility is that a vector containing the gene of mutated β-catenin is transfected into a wild-type cell. In the theoretical analyses we consider the case of transfection. Hence, the production of wild-type β-catenin is equal in the wild-type and the mutated model. In the case of the mutation of one allele, the production of wild-type β-catenin in the mutated system would be lowered compared to the wild-type system. Thus, the absolute concentrations are different in the transfected and the mutated case. However, the overall observations are the same (comparison not shown).

In the following sections, the results of investigations with respect to the effects on the steady state concentrations and dynamical behaviour are shown. Analyses concerning the effects on the fold-change are shown in Appendix B.3.

2.3.2   Effect of mutated β-catenin on the unstimulated steady state concentration of wild-type β-catenin

The unstimulated system is in a steady state. The steady state concentrations are determined by the kinetic parameters. Here, we are interested in how the unstimulated steady state concentration of wild-type β-catenin is influenced by the presence of mutated β-catenin and whether different steady state concentrations of the mutated protein affect the concentration of the wild-type protein. For this analysis, we calculate the unstimulated steady state concentration of wild-type β-catenin for varying production rates of the mutated protein. The production rate of mutated β-catenin is altered as a multiple of the production rate of wild-type β-catenin that always keeps its reference value. In Figure 2.13, the steady state of wild-type β-catenin is plotted against the ratio f of the production rates

$$f = \frac{\text{production of mutated } \beta\text{-catenin}}{\text{production of wild-type } \beta\text{-catenin}} \qquad\qquad (2.5)$$

that reflects the factor of increase of the production rate of mutated β-catenin compared to the production rate of the wild-type protein.

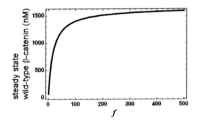

Figure 2.13: Dependence of the unstimulated steady state concentration of wild-type β-catenin on the production rate of the mutated form.

The ratio f gives the ratio of the production rates of mutated and wild-type β-catenin. The production rate of the wild-type protein keeps its reference value.

Figure 2.13 illustrates that the unstimulated steady state concentration of wild-type β-catenin increases with an increasing production rate of the mutated form. The more mutated β-catenin exists, the higher is the unstimulated steady state level of wild-type β-catenin. The underlying reason is that mutated β-catenin competes with the wild-type protein for the binding partners APC, TCF and the destruction complex APC*/Axin*/GSK3. The more mutated protein is available, the more wild-type β-catenin is displaced from the complexes, leading to a decreased degradation of wild-type β-catenin. This effect is strongly enhanced by the sequestration of a high amount of APC by mutated β-catenin. Due to this, the availability of APC for the formation of the destruction complex is strongly decreased. This effect is discussed in the context of APC mutations in section 2.2. Taken together, as a consequence of the presence of mutated β-catenin, the concentration of unbound wild-type β-catenin increases as less wild-type β-catenin is degraded via the destruction cycle. With increasing values of f, the curve flattens and the unstimulated steady state concentration of wild-type β-catenin saturates. This saturation concentration is determined by the production rate of wild-type β-catenin (k_{12}) and the rate of its alternative degradation (k_{13}). If the mutated protein exists in a very high concentration, it occupies all binding partners. Under these circumstances, the concentration of wild-type β-catenin is exclusively determined by its turnover, reaching its maximal value of ~1646 nM.

From this analysis we conclude that in the presence of mutated β-catenin the unstimulated steady state concentration of wild-type β-catenin increases. Furthermore it yields that the more mutated β-catenin exists, the higher the steady state concentration of the wild-type protein until it reaches its maximal steady state concentration determined by its production and alternative degradation.

Experimental confirmation of the theoretical prediction

The experiments for the verification of the theoretical prediction that the presence of mutated β-catenin leads to an increase of the steady state concentration of wild-type β-catenin have been performed in Rolf Gebhardt's group at the University Leipzig. For this purpose they have used Huh7 hepatoma cells, a human cell line. These cells contain wild-type β-catenin exclusively. They have been transfected with a vector containing the gene of mutated β-catenin lacking exon-3. 24 hours after transfection, wild-type and mutated β-catenin have been detected by Western blot and analysed densitometrically. In both cases, the relative density changes have been measured. In transfected cells, wild-type as well as mutated β-catenin have been determined. In untransfected wild-type cells, wild-type β-catenin has been measured as control. The results are shown in Figure 2.14.

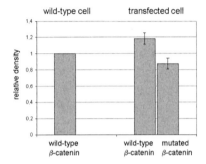

Figure 2.14: Wild-type and mutated β-catenin levels in transfected and non-transfected Huh7 cells.

Huh7 cells have been transfected with a vector containing mutated β-catenin lacking exon-3. After 24 h, wild-type and mutated β-catenin have been measured by Western blot. The relative density changes detected in three experiments are plotted with the standard deviation. The level of wild-type β-catenin in untransfected cells is used as control. For each experiment, the control is set to one.

In Figure 2.14 it is demonstrated that the transfection with a vector containing mutated β-catenin leads to an increase of the wild-type β-catenin concentration. Compared to the control, the concentration of wild-type β-catenin is increased by 18% ± 7% if the cell is transfected with mutated β-catenin. The increase is in full accordance with the theoretical prediction.

2.3.3   Stimulated system

Dependence of the stimulated steady state levels of the two β-catenin forms on the stimulation strength

Next, we are interested in the effects of different stimulation strengths on the stimulated steady state level of wild-type and mutated β-catenin. In the corresponding experiments, the pathway is activated by the inhibition of the GSK3 upon application of LiCl. This is a common method to activate the Wnt/β-catenin pathway. The stronger the GSK3 is inhibited, the stronger the pathway is activated. To match these experiments, the stimulus is mimicked by an inhibition of the GSK3 in the simulations. This is realised by reducing the rate constants of the phosphorylation reactions mediated by GSK3. These reactions are the phosphorylation of APC and Axin bound to GSK3 (reaction 4) and the phosphorylation of β-catenin bound to the destruction complex (reaction 9). It is assumed that the GSK3 inhibitor has the same percentaged effect on both reactions. Figure 2.15 shows the dependence of the inhibition strength on (A) the level of wild-type β-catenin and (B) the concentration of the mutated protein.

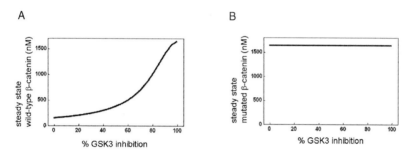

Figure 2.15: Dependence of the stimulated steady state of wild-type and mutated β-catenin on the strength of the stimulation.

The stimulated steady state concentration of (A) wild-type and (B) mutated β-catenin are calculated. The values are plotted against the strength of GSK3 inhibition. The inhibition of the GSK3 mimics the pathway stimulation. This is realised by altering the rate constants of the two reactions describing the phosphorylation steps mediated by GSK3 (reactions 4 and 9) by equal percentaged amounts.

Figure 2.15A reveals that the stimulated steady state concentration of wild-type β-catenin increases with an increasing activation of the pathway. This relation is non-linear. If the GSK3 is inhibited by ~74%, the half maximal wild-type β-catenin concentration is reached. For low pathway activations, an increase (e.g. from 10% to 20%) has a smaller effect on the wild-type β-catenin concentration than for higher stimulation strengths (e.g. from 70% to 80%). If the pathway is slightly activated by GSK3 inhibition, the destruction cycle still works very efficiently and a high amount of β-catenin is degraded via this mechanism. If the GSK3 is strongly inhibited, less β-catenin is degraded via the destruction cycle and its stimulated steady state level increases. For a complete inhibition, the β-catenin concentration reaches its maximal value determined by its production (k_{12}) and alternative degradation rate (k_{13}). This is equal to the steady state of mutated β-catenin. Over the whole inhibition range, the steady state concentration of mutated β-catenin remains constant on the maximal level of

~1646 nM that is only determined by its own turnover (Figure 2.15B). With the exception of the GSK3 inhibition by 100%, the stimulated steady state concentration of mutated β-catenin is higher than the steady state concentration of wild-type β-catenin.

Based on the simulations we predict distinct dependences of wild-type and mutated β-catenin on increasing pathway activations in a system containing both protein forms. The concentration of mutated β-catenin is higher than the concentration of the wild-type protein. With increasing strength of the pathway stimulation, the concentration of wild-type β-catenin increases while the concentration of mutated β-catenin stays constant.

Experimental confirmation of the theoretical predictions

The theoretical prediction that an increasing stimulus strength causes higher levels of wild-type β-catenin but will not affect the level of mutated β-catenin, is verified experimentally using HepG2 cells. These are human liver carcinoma cells in which wild-type and mutated β-catenin coexist. Hence, no additional transfection with vectors containing mutated β-catenin is necessary. The HepG2 cells are treated with LiCl, an inhibitor of the GSK3. The concentrations of wild-type and mutated β-catenin are determined for different LiCl concentrations (Figure 2.16).

A
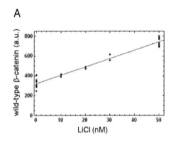

B
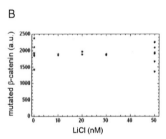

Figure 2.16: Concentration of wild-type and mutated β-catenin in HepG2 cells treated with LiCl.

This analysis is done using HepG2 cells which contain both wild-type and mutated β-catenin. The cells are treated with LiCl (0 nM, 10 nM, 20 nM, 30 nM, 50 nM). 8 h after stimulation, (A) wild-type as well as (B) mutated β-catenin are determined by Western blot. The densities of the bands are detected and plotted against the respective LiCl concentration. The densities are indicated as arbitrary units (a.u.).

In (A), the linear regression line is plotted.

The experimental results shown in Figure 2.16 demonstrate that increasing LiCl concentrations have distinct effects on the concentration of wild-type and mutated β-catenin. In agreement with the model predictions, an increase in the LiCl concentration does not lead to a change in the stimulated steady state concentration of mutated β-catenin (Figure 2.16B). In contrast, the stimulated steady state of wild-type β-catenin increases with increasing LiCl concentrations (Figure 2.16A). Within the considered concentration range, the LiCl concentration and the concentration of wild-type β-catenin are linearly related ($R^2 = 0.95$).

The calculations (Figure 2.15A) do not predict a linear relation between wild-type β-catenin and the GSK3 inhibition over the whole range of GSK3 inhibition. This deviation between model prediction and experimental observation indicates that with the chosen LiCl concentrations only a smaller range of GSK3 inhibition is covered. So far, no information has been published on the extent of GSK3 inhibition by LiCl. Another explanation is that the GSK3 inhibition might also affect further cellular functions of the GSK3 (Doble and Woodgett, 2003) which are not included in the model.

The temporal behaviour of wild-type and mutated β-catenin upon pathway stimulation

As a next step, we consider the situation of constant pathway activation by the inhibition of the GSK3 by 70%. To this end, the rate constants of the phosphorylation reactions mediated by GSK3 (reactions 4 and 9) are diminished by 70%. The temporal behaviour of β-catenin is similar in the simulations in which the pathway is activated by the GSK3 inhibition and in the simulations in which the pathway is activated by a Wnt stimulus (comparison not shown). Figure 2.17 shows the temporal behaviour of (A) wild-type and (B) mutated β-catenin. For comparison, the temporal behaviour of β-catenin in the wild-type model is presented (dotted line in Figure 2.17A).

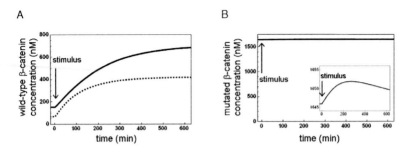

Figure 2.17: Temporal behaviour of wild-type β-catenin and mutated β-catenin.
The pathway is constantly stimulated by inhibiting the GSK3 by 70% starting at $t = 0$ min.
In (A), the dynamics of wild-type β-catenin are shown in both the wild-type (dotted line) and
the mutated model (solid line). In (B), the temporal behaviour of mutated β-catenin is
presented. The inset shows a smaller concentration range.

Comparing the behaviour of wild-type β-catenin in the absence and presence of the mutated
protein (Figure 2.17A) shows that the steady states are increased if the mutated protein is
present (see also section 2.3.2). The unstimulated steady state concentration of wild-type β-
catenin rises by a factor of ~2.24 from 68.7 nM to 153.8 nM. Mutated β-catenin sequesters
binding partners of β-catenin, especially APC, leading to an increase in the concentration of
wild-type β-catenin. The ratio between the stimulated and the unstimulated steady state levels
decreases if the mutated protein form is available. The ratio declines from ~3.37 to ~2.93 (see
also Figure B.1, Appendix B). Taken together, in the mutated model the β-catenin steady
states are increased and the fold-change is decreased compared to the corresponding
characteristics in wild-type model. This is similar to the effects caused by an APC mutation
(see section 2.2.3). Compared to the wild-type model, the stimulated steady state is reached
later in the presence of mutated β-catenin. In the mutated model, 99.5% of the stimulated
steady state concentration is reached at ~1023 min while in the wild-type model, this level is
already reached at ~675 min.

It can be seen that upon stimulation the concentrations of wild-type and mutated β-catenin
protein behave differently in the mutated model (compare solid lines in Figure 2.17A and
Figure 2.17B). The concentration of wild-type β-catenin increases to a higher steady state
level (Figure 2.17A). The concentration of the mutated protein transiently increases
marginally before it reaches its original steady state concentration again (Figure 2.17B). The
maximal concentration change is less than 1%. The level of mutated β-catenin remains higher

than the steady state level of the wild-type protein since mutated β-catenin is only degraded via the destruction-cycle-independent mechanism. The transient concentration change of the mutated protein is caused by the fact that the behaviour of wild-type and mutated β-catenin are mutually dependent. Since the concentration of wild-type β-catenin increases, more wild-type protein is available and displaces the mutated protein from the complexes. This leads to an increase in the concentration of unbound mutated β-catenin. The degradation of mutated unbound β-catenin increases and its concentration declines again. After some time, the level of the unstimulated steady state concentration is reached again. The temporary change in the concentration of mutated β-catenin is very small as only a small amount of mutated β-catenin is replaced from the complexes. Experimentally, it might be impossible to detect the concentration change of mutated β-catenin.

In essence, we conclude that a stimulation of cells containing both β-catenin forms should lead to a detectable concentration change of the wild-type protein over time. In comparison to the wild-type model, the increase of the β-catenin concentration takes longer if mutated β-catenin is additionally available in the system. The concentration of the mutated protein remains almost constant on a higher level than the wild-type protein. For the steady state, this is in agreement with the experimental observations shown in Figure 2.15. Time course data of wild-type and mutated β-catenin has yet not been measured.

2.3.4   Sensitivity analysis

The dynamics of the system are to some extent determined by the model's kinetic parameters. Sensitivity coefficients are used to analyse the effects of the system behaviour due to parameter changes. The coefficients describe the relative changes of a specific property with respect to relative changes of the rate constants. Here, the unstimulated steady state concentration of wild-type β-catenin is chosen as specific characteristic (*characteristic*). The sensitivity coefficients are defined by

$$C_i^{characteristic} = \frac{k_i}{characteristic} \frac{\mathrm{d}(characteristic)}{\mathrm{d}(k_i)} \tag{2.6}$$

with $C_i^{characteristic}$ being the sensitivity coefficient regarding parameter k_i.

The sensitivity coefficients are presented as bar plots. A positive sensitivity coefficient means that an increase in the respective parameter leads to an increase of the unstimulated steady state level of β-catenin.

For very small perturbations, sensitivity coefficients are equal to control coefficients. The concept of control coefficients was originally established for analysing the control in metabolic networks (Heinrich et al., 1978; Kacser and Burns, 1973) but is nowadays also applied to the analyses of signalling pathways.

Sensitivity coefficients of the unstimulated steady state concentration of wild-type β-catenin

In the next theoretical investigation, the impact of the reactions on the unstimulated steady state of wild-type β-catenin is analysed. Each parameter is changed by 1% and the sensitivity coefficients are calculated (Figure 2.18).

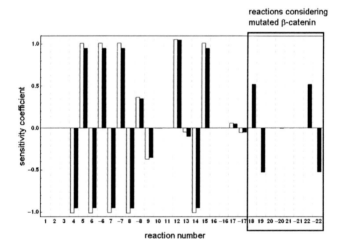

Figure 2.18: Sensitivity coefficients of the unstimulated steady state concentration of wild-type β-catenin.

The sensitivity coefficients with respect to the unstimulated steady state of wild-type β-catenin are calculated in the wild-type model (white bars) and the mutated model (black bars). The parameters are changed by 1%. The reactions dealing with mutated β-catenin (reactions 18 to –22) are framed in blue.

The numbers correspond to the reaction numbers shown in the model scheme in Figure 2.12. In the case of reversible reactions, the complex formation is denoted with a positive sign; the dissociation of the complex is labelled with the negative reaction number.

Figure 2.18 illustrates the impact of the individual processes on the β-catenin steady state changes in the presence of mutated β-catenin. In the wild-type model, the main positive regulation of the concentration of wild-type β-catenin occurs through i) the processes leading to the decomposition of the destruction complex (reactions 5, –6, –7, and –8), ii) the production of β-catenin (reaction 12), and iii) the degradation of Axin (reaction 15). The concentration is regulated negatively by i) the processes facilitating the formation of the destruction complex and the passing through the destruction cycle (reactions 4, 6, 7, 8 and 9), ii) the alternative β-catenin degradation (reaction 13), and iii) the production of Axin (reaction 14), see white bars in Figure 2.18.

In the mutated model (black bars in Figure 2.18), reactions that take mutated β-catenin into consideration have an impact on the unstimulated steady state of wild-type β-catenin. This confirms the analysis shown in section 2.3.2. The sensitivity analysis reveals a strong control of the synthesis and degradation of mutated β-catenin (reactions 18 and 19, respectively) as well as the formation and dissociation of the APC/β-catenin^{mut} complex (reaction 22). In contrast, the formation and dissociation of the TCF/β-catenin^{mut} complex (reaction 21) and the formation and dissociation of the APC*/Axin*/GSK3/β-catenin^{mut} complex (reaction 20) have almost no impact on the unstimulated steady state of wild-type β-catenin. Hence, the main effect on this steady state mediated by mutated β-catenin is to occupy APC, leading to a lower availability of APC for the composition of the destruction complex. The sequestration of the destruction complex APC*/Axin*/GSK3 by mutated β-catenin (reaction 20) has only a very low impact.

In the mutated model, reactions taking wild-type β-catenin into account (reactions 1 to –17) mainly affect the unstimulated steady state of wild-type β-catenin. Comparing the coefficients of these reactions in the wild-type model and the mutated model reveals that the impact of the individual reactions is very similar independent of the presence or absence of mutated β-catenin. In the presence of mutated β-catenin, the impact of the single reactions decreases slightly. The only exception of this general observation is the impact of the alternative β-catenin degradation (reaction 13) that increases marginally if mutated β-catenin is available. The reason is that in the presence of mutated β-catenin a lower amount of wild-type β-catenin is degraded via the destruction cycle and more β-catenin is degraded via the alternative degradation. Hence, its impact increases. The increase of the impact of the alternative β-catenin degradation is also observed if APC is mutated. Under these circumstances, also a lower amount of β-catenin is degraded via the destruction cycle (see section 2.2.3).

Interestingly, some processes have a distinct effect depending on whether the process is related to wild-type or mutated β-catenin. In the following, these processes are referred to as analogue processes. The considered reactions are the interaction of the particular β-catenin forms with i) the complex APC*/Axin*/GSK3 (reactions 8 and 20), ii) TCF (reactions 16 and 21) as well as iii) APC (reactions 17 and 22). Furthermore, iv) the alternative degradations (reactions 13 and 19) and v) the productions (reactions 12 and 18) of wild-type and mutated β-catenin are taken into account. The effect is illustrated in Figure 2.19 which shows the ratio of the sensitivity coefficients of analogue reactions. The sensitivity coefficient of a reaction related to mutated β-catenin is divided by the sensitivity coefficient of the analogue process dealing with wild-type β-catenin.

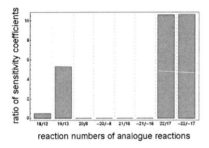

reaction numbers of analogue reactions

Figure 2.19: Ratio of the sensitivity coefficients of analogue reactions.
The sensitivity coefficient of a reaction related to mutated β-catenin is divided by the sensitivity coefficient of the analogue process dealing with wild-type β-catenin. The sensitivity coefficients of the unstimulated steady state of wild-type β-catenin are considered. The numbers correspond to the reactions numbers shown in the model scheme in Figure 2.12. In the case of reversible reactions, the complex formation is denoted with a positive sign; the dissociation of the complex is labelled with the negative reaction number.

The analysis of this ratio emphasises the finding that analogue reactions have impacts of different strength on the unstimulated steady state of wild-type β-catenin (Figure 2.19). Analogue reactions that affect the steady state of wild-type β-catenin are: i) the synthesis reactions (reactions 12 and 18), ii) the alternative degradations (reactions 13 and 19), and iii) the formation and dissociation of the APC/β-catenin and the APC/β-catenin^{mut} complex (reactions 17 and 22). The impact of the interaction of mutated β-catenin with TCF (process 21) and the destruction complex (process 20) are very small, and the calculated ratios are very small.

All ratios are positive. Independent of whether the processes are related to wild-type or mutated β-catenin, the type of effect on the wild-type β-catenin steady state is the same. This means that for instance the inhibition of the alternative degradation of wild-type β-catenin (reaction 13) causes a lower steady state and the inhibition of the alternative degradation of mutated β-catenin (reaction 19) results in a lower wild-type β-catenin steady state as well.

The ratio of the sensitivity coefficient of the production rates (18/12) is less than one (~0.5). The impact of the production of wild-type β-catenin has a higher impact on its steady state than the production of the mutated protein. In contrast, the ratios concerning the alternative degradation (19/13) and the interaction with APC (22/17 and −22/−17) are greater than one. The unstimulated steady state of wild-type β-catenin is stronger affected by the reactions related to mutated β-catenin than by the reactions related to wild-type β-catenin.

The analyses shown in Figure 2.18 and Figure 2.19 lead to the interpretation that in the mutated model the concentration of wild-type β-catenin is largely controlled by the reactions dealing with wild-type β-catenin (reactions 1 to −17). The processes related to mutated β-catenin are of high importance by binding of mutant β-catenin to APC (process 22) resulting in a reduction of APC available for the wild-type processes. Furthermore, the production and alternative degradation of mutated β-catenin (reactions 18 and 19, respectively) influence the concentration of wild-type β-catenin as they regulate the availability of mutated β-catenin.

Prediction for interventions

Based on the sensitivity analysis, useful interventions to diminish the effects of mutated β-catenin on the unstimulated steady state level of wild-type β-catenin can be predicted. Due to the mutated protein the concentration of wild-type β-catenin is increased without the pathway stimulation.

As already argued in section 2.2, for medical applications, the most interesting inhibitors would be those that have strong effects on the wild-type β-catenin steady state level in the mutated model, but that only weakly affect that steady state in the wild-type model. In other words: The steady state level in a healthy cell should be altered only slightly while it should be changed strongly in the abnormal cell. The treatment with such an inhibitor would only have weak side effects in healthy tissues.

The sensitivity analysis reveals that the main impact of mutated β-catenin is mediated by its ability to sequester APC in the APC/β-catenin^{mut} complex (reaction 22). Additionally, the analogue process using wild-type β-catenin (reaction 17) has only a very small impact in both the wild-type and the mutated model. Hence, the inhibition of the binding of APC and β-catenin (wild-type and mutated β-catenin) fulfils the requests for a perturbation that affects the mutated model strongly but the wild-type model only marginally. The effects of the inhibition of these complex formations (reactions 17 and 22) on the wild-type β-catenin steady state and dynamics are shown in Figure 2.20. In addition, the effects of an APC overexpression are simulated. The main effect of mutated β-catenin is to reduce the amount of APC that is available for the formation of the destruction complex. An overexpression counteracts this effect and might be easier to realise in experiments than a selective inhibition of the binding of APC and wild-type and mutated β-catenin (reactions 17 and 22, respectively).

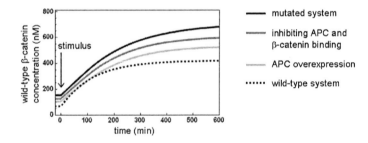

Figure 2.20: Effects of interventions on the dynamics of wild-type β-catenin.
The dashed black line shows the β-catenin dynamics in the wild-type model while the solid black line represents the temporal behaviour of wild-type β-catenin in the mutated model. In the mutated model, two interventions are examined: i) The inhibition of the binding of APC and β-catenin (processes 17 and 22) by 30% (dark grey line), and ii) an overexpression of APC (total amount of APC is 150 nM instead of 100 nM; light grey line). The constant stimulus is given at $t = 0$ min. It is implemented as the inhibition of the GSK3 by 70%.

The effects of the interventions on the β-catenin steady states and dynamics are shown in Figure 2.20. They reveal that both the inhibition of the binding of β-catenin and APC (reactions 17 and 22; dark grey line) and the overexpression of APC (light grey line) result in decreased steady states of wild-type β-catenin in comparison to the level in the unperturbed mutated model (solid black line). The unstimulated and the stimulated steady state levels are diminished. Also the duration of the increase of the β-catenin concentration upon pathway activation is affected by the interventions. 99.5% of the stimulated steady state concentrations of wild-type β-catenin are reached earlier than in the mutated model in which this increasing phase lasts ~1023 min but still later than in the wild-type model (~675 min). If APC is overexpressed, the level is reached at ~765 min while the increasing phase lasts ~835 min if the binding of APC and β-catenin is inhibited.

The inhibition of the binding of APC and β-catenin (reactions 17 and 22) results in a higher level of APC that is available for the formation of the destruction complex. Consequently, a higher amount of wild-type β-catenin is degraded via the destruction cycle. The overexpression of APC also results in a higher availability of the destruction complex, and thus, a lower wild-type β-catenin concentration. However, such an overexpression of APC diminishes the wild-type β-catenin level also in the wild-type model (not shown). Therefore, a tissue-wide APC overexpression would cause strong side effects in healthy cells. Compared to the wild-type model (dashed line), the β-catenin steady states are still increased under both cases of interventions. Stronger inhibitions or overexpressions are necessary to decrease the concentration to such an extent that it is more similar to the levels in the wild-type model (dotted line). However, an exact match of the dynamics of wild-type β-catenin in the intervened mutated model and the wild-type model is not possible. In the mutated model, an APC concentration of 230 nM decreases the unstimulated wild-type β-catenin steady state to ~68.8 nM which is very similar to the level in the wild-type model (~68.7 nM). However, the stimulated steady state concentrations differ considerably; in the wild-type model, this level is ~422 nM while in the intervened mutated model a steady state concentration of ~378 nM is reached. Also the inhibition of the binding of APC and β-catenin (reactions 17 and 22) in the mutated model that results in an unstimulated steady state level that is similar to the level in the wild-type model causes a stimulated steady state level that is much lower than the corresponding level in the wild-type model.

Experimentally it has been shown that the transcriptional activity in HepG2 cells, in which wild-type and mutated β-catenin coexist, has been decreased if the cells have been transfected with APC (Satoh et al., 2000). This corresponds with the simulation results.

We conclude that an inhibitor affecting selectively the binding of β-catenin and APC would be a potent drug to antagonise the increase of the level of the wild-type β-catenin caused by the presence of mutated β-catenin. Until today, it is impossible to inhibit this interaction selectively in experiments. Recent studies have focussed on the domains mediating the binding of APC and β-catenin (Kohler et al., 2010; Kohler et al., 2009; Kohler et al., 2008; Munemitsu et al., 1995; Yang et al., 2006). New details about these domains might promote the development of drugs that affect this interaction.

One has to keep in mind that the analysed perturbations only aim at the reduction of the level of wild-type β-catenin. There is still a high amount of mutant protein in the system which may induce additional, yet unknown processes. For instance this could be the regulation of the expression of genes that are not targeted by wild-type β-catenin.

2.3.5   Minimal model

In the future, the effects of the coexistence of wild-type and mutated β-catenin on the expression of target genes will be analysed. To this end, we develop a minimal model that describes the β-catenin dynamics qualitatively and reflects the main dependences and interrelations of wild-type and mutated β-catenin. One advantage of the minimal models over the detailed models is that in part, a minimal model can be analysed analytically. This allows for general conclusions. For the purpose of comparison, a wild-type minimal model and a mutated minimal model are developed. In the following, the wild-type and mutated models analysed in sections 2.3.2 to 2.3.4 are referred to as detailed wild-type and detailed mutated model. Like the detailed models, the minimal wild-type model and the minimal mutated model are compared to each other. Furthermore, the behaviour observed in the minimal models is compared with the behaviour observed in the detailed models. The models introduced here are developed with respect to the analyses of the effects of mutated β-catenin on the steady state and dynamics of the wild-type protein. In this respect they differ from a previously developed minimal model (Mirams et al., 2010). The model reduction is based on the knowledge about the impact of reactions and components on wild-type β-catenin.

Model scheme

In Figure 2.21, a scheme of the developed minimal models is shown. The wild-type model is shown in black, the additional reactions taking place in the mutated model are coloured in blue. The minimal wild-type model consists of the following reactions: wild-type β-catenin is produced and degraded (reactions 1 and 2, respectively). Reaction 2 reflects the alternative degradation of β-catenin. Furthermore, the wild-type protein is degraded in an APC-dependent way (reaction 3), reflecting the β-catenin degradation via the destruction cycle. APC reversibly forms a complex with β-catenin (reaction 4). In the mutated system, the following reactions are included in addition: mutated β-catenin (β-catenin^{mut}) is produced (reaction 5) and degraded (reaction 6). Like the wild-type protein, mutated β-catenin reversibly binds APC (reaction 7). The stimulus is simulated as a decrease of the rate constant of reaction 3. It reduces the degradation of wild-type β-catenin via the destruction cycle.

The minimal models concentrate on the interplay of wild-type and mutated β-catenin via their competition to bind APC. This varies the availability of APC. The analyses of the detailed models reveal that the main impact of mutated β-catenin on the wild-type protein is mediated via the sequestration of APC by mutated β-catenin (see section 2.3.4).

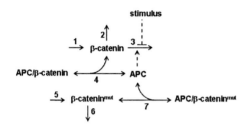

Figure 2.21: Reaction scheme of the minimal wild-type and minimal mutated model of the Wnt/β-catenin signalling pathway.

Reactions and components of the minimal wild-type model are shown in black. The reactions and components occurring in addition in the minimal mutated model are coloured in blue.

Numbers next to arrows denote the number of the particular reaction. One-headed arrows denote reactions taking place in the indicated direction, dashed arrows represent activation steps and double-headed arrows illustrate binding reactions. The reversed, dashed T denotes the inhibition of reaction 3 by the stimulus.

In the minimal models, reactions are merged compared to the detailed models (Figure 2.12). For example, all reactions involved in the destruction cycle (reactions 8 to 10 in Figure 2.12) are merged into reaction 3 in the minimal model. The production rates and alternative degradations of wild-type and mutated β-catenin (reactions 12, 13, 18 and 19 in the detailed model) correspond to the reactions 1 and 2 (wild-type β-catenin) as well as processes 5 and 6 (β-catenin^{mut}) in the minimal model. The reactions 4 and 7 correspond to the binding reactions to APC in the detailed model (reactions 17 and 22). Further reactions are neglected. In addition, not all components of the detailed models are considered such as TCF. Some components are integrated implicitly like GSK3 that is involved indirectly via reaction 3.

By these simplifications the wild-type model is reduced to three variables and four reactions (compared to 15 variables and 17 reactions in the detailed wild-type model, Figure 2.12) and for the minimal mutated system to five components and seven reactions (compared to 19 variables and 22 reactions in the detailed mutated model, Figure 2.12).

The model equations, kinetics and parameters are provided in Appendix B. The minimal wild-type model is given by equations B.15 to B.17 and B.21 to B.25; the minimal mutated model is given by the equations B.15, B.17 and B.18 to B.29. As we are just interested in the general qualitative behaviour, the kinetic parameters are not adjusted (Table 9). They are given in time units (tu) and concentration units (cu). Like in the detailed model, parameters of analogue reactions have the same values independent of whether they are related to wild-type or mutated β-catenin.

β-Catenin dynamics in the minimal models

It is examined whether the minimal models are in general able to reproduce the same features as the detailed models. First, we investigate whether the minimal models enable the reproduction of the typical temporal behaviour of wild-type and mutated β-catenin upon constant pathway activation. The dynamics of mutated and wild-type β-catenin in the wild-type and mutated minimal models are shown in Figure 2.22.

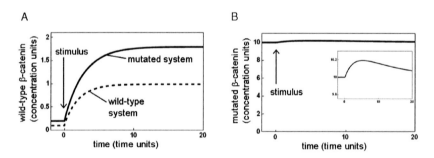

Figure 2.22: Temporal behaviour of wild-type and mutated β-catenin in the minimal models. (A) The behaviour of wild-type β-catenin is shown. The dynamics in the minimal wild-type model (dashed line) as well as the dynamics in the minimal mutated model (solid line) are presented. (B) The temporal behaviour of mutated β-catenin is shown. The inset focuses on a smaller concentration range. The constant stimulus is given at $t = 0$ tu and modelled as a switch from $k_3 = 1$ cu^{-1}tu^{-1} to $k_3^{stim} = 0.1$ cu^{-1}tu^{-1}.

The simulations of the minimal models reveal that the concentration of wild-type β-catenin is higher in the minimal mutated model than in the minimal wild-type model (compare solid and dashed line in Figure 2.22A). In the minimal wild-type model, the unstimulated steady state of β-catenin is ~0.099 cu. In the minimal mutated model, the unstimulated steady state level is about 0.198 cu. The stimulated steady state levels are ~0.99 cu and ~1.789 cu in the minimal wild-type and minimal mutated model, respectively. Thus, the fold-change in the minimal wild-type model is ~10 and higher than in the minimal mutated model in which the fold-change is ~9. In the minimal mutated model, the level of mutated β-catenin is higher than the level of wild-type β-catenin (Figure 2.22B). As in the detailed model, the concentration of mutated β-catenin transiently increases, but returns to its original steady state level. The transient concentration change is very small. These general observations are in agreement with the findings in the detailed models (Figure 2.17). Hence, the minimal models reproduce very well the general temporal behaviour of wild-type and mutated β-catenin observed in the detailed wild-type and mutated models.

Dependence of the unstimulated steady state concentration of wild-type β-catenin on the production rate of the mutated protein

The dependence of the steady state concentration of wild-type β-catenin on the production rate of mutated β-catenin is presented in Figure 2.23. This corresponds to the analysis in the detailed mutated model shown in Figure 2.13. As for the detailed mutated model, the unstimulated steady state of wild-type β-catenin is plotted versus the ratio f which is the ratio of the production rates of mutated and wild-type β-catenin (equation 2.5, section 2.3.2).

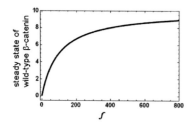

Figure 2.23: Dependence of the unstimulated steady state of wild-type β-catenin on the production rate of mutated β-catenin.

The ratio f gives the ratio of the production rates of mutated and wild-type β-catenin (equation 2.5, section 2.3.2). The rate of the wild-type protein keeps its reference value.

The analysis shown in Figure 2.23 reveals that in the mutated minimal model the dependence of the unstimulated steady state concentration of wild-type β-catenin on the production rate of mutated β-catenin is similar to the dependence observed in the detailed mutated model (Figure 2.13). The more mutated β-catenin is produced, the higher the unstimulated steady state level of wild-type β-catenin. For very high production rates of the mutated protein, the concentration of wild-type β-catenin reaches its maximal value which is determined by its own production and alternative degradation. This relation is also valid in the detailed model (see section 2.3.2). Thus, this property is also covered by the minimal mutated model.

The explicit dependence of the steady state of wild-type β-catenin on the parameters is analysed analytically below in this section (paragraph "Analytical calculation of the steady state of wild-type β-catenin").

Sensitivity coefficients

As a next step, we determine the sensitivity coefficients of the unstimulated steady state of wild-type β-catenin (shown in Figure 2.24). The coefficients are calculated as described in section 2.3.4, equation 2.6. They are compared to the coefficients calculated for the detailed models, shown in Figure 2.18.

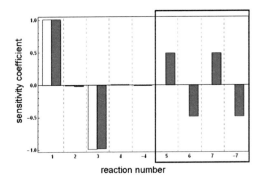

Figure 2.24: Sensitivity coefficients of the unstimulated steady state level of wild-type β-catenin in the minimal wild-type and minimal mutated model.

For the calculation of the coefficients the parameters are changed by 1%. The white and grey bars are related to the minimal wild-type and the minimal mutated model, respectively. The reactions that take mutated β-catenin into account are framed in blue. The numbers correspond to the reactions numbers shown in the model scheme in Figure 2.21. In the case of reversible reactions, the complex formation is denoted with a positive sign; the dissociation of the complex is labelled with the negative reaction number.

The analysis shows that processes which are comparable in the minimal and the detailed models (see paragraph "Model scheme") have similar impacts on the unstimulated steady state of wild-type β-catenin (compare Figure 2.18 and Figure 2.24). In general, the reactions involving mutated β-catenin (reactions 5 to 7; framed in blue) affect the concentration of wild-type β-catenin in a similar way as in the detailed mutated model. The impact of the binding of mutated β-catenin and APC (reaction 7) is stronger than the impact of the binding of wild-type β-catenin and APC (reaction 4). The impact of the production of wild-type β-catenin (reaction 1) is high as well as its degradation which is activated by APC (reaction 3). The impact of the alternative degradation of wild-type β-catenin (reaction 2) is very small but slightly increases if mutated β-catenin is present. This impact is still much lower than the impact of the degradation of mutated β-catenin (reaction 6). Comparable statements are drawn based on the analyses of the detailed models (section 2.3.4).

Due to the reduction, some features of the detailed models are not reproducible by the minimal models, for instance the different influences of the binding of mutated β-catenin and APC or the destruction complex.

Interventions

Moreover, we investigate whether the effects of interventions analysed in the detailed mutated model (Figure 2.20) have comparable effects in the mutated minimal model (Figure 2.25). The wild-type β-catenin dynamics in the minimal wild-type and minimal mutated models are compared to the temporal behaviour in the intervened mutated minimal model. The considered perturbations are: i) the inhibition of the binding of APC and wild-type and mutated β-catenin as well as ii) the overexpression of APC.

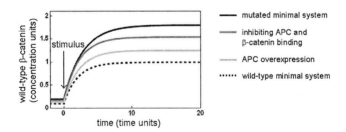

Figure 2.25: Effects of the perturbations on the dynamics of wild-type β-catenin in the minimal mutated model.

The solid black line shows the wild-type β-catenin dynamics in the minimal mutated model. For comparison, the β-catenin time course in the minimal wild-type model is shown (dotted line). In dark grey, the β-catenin dynamics in the mutated minimal model is shown if the binding of APC and β-catenin is inhibited by 30%. In light grey, the dynamics of wild-type β-catenin is shown if a higher amount of APC is available (total $APC = 15$ cu instead of total $APC = 10$ cu).

Figure 2.25 illustrates that the perturbations affect the β-catenin dynamics in the minimal mutated model in a similar way as shown in the detailed mutated model (for comparison see Figure 2.20). The inhibition of the binding of APC and the two β-catenin forms (reactions 4 and 7) as well as a higher APC level reduce the concentration of wild-type β-catenin in the mutated minimal model. Under these conditions, the difference between the dynamics of wild-type β-catenin in the mutated and wild-type minimal model is diminished. This is in agreement with the results obtained by the analysis of the detailed model (Figure 2.20).

In summary, the minimal models represent the detailed models very well. They show comparable dynamics of the pathway components and properties of regulation.

Analytical calculation of the steady state of wild-type β-catenin

The minimal model enables the analytical calculation of the steady state. We are interested in the steady state level of wild-type β-catenin and a comparison of this steady state level in the minimal wild-type and the minimal mutated model.

In the minimal wild-type model, the steady state of wild-type β-catenin (x_w) is given by:

$$x_w = \frac{1}{c_1}(c_2 - c_3 - c_4 \cdot A) + \frac{1}{c_1} \cdot \sqrt{(c_2 + c_3)^2 + (c_4 \cdot A)^2 + 2 \cdot c_4 \cdot A \cdot (-c_2 + c_3)}, \qquad (2.7)$$

with

$c_1 = 2 \cdot k_2 k_4$,

$c_2 = k_1 k_4$,

$c_3 = k_2 k_{-4}$, and

$c_4 = k_3 k_{-4}$.

All k_i are positive. Therefore it holds that $c_1, c_2, c_3, c_4 > 0$. A is the total concentration of APC, with $A > 0$. The wild-type β-catenin steady state depends on all rate constants of the minimal wild-type model and the total concentration of APC.

There exists always a real solution for x_w, because the inequation

$$(c_2 + c_3)^2 + (c_4 \cdot A)^2 + 2 \cdot c_4 \cdot A \cdot (-c_2 + c_3) > 0 \qquad (2.8)$$

is always fulfilled, for all A, c_2, c_3, $c_4 > 0$. This can be seen easily after the transformation of the inequation to:

$$(c_2 - c_4 \cdot A)^2 + c_3^2 + 2 \cdot c_2 \cdot c_3 + 2 \cdot A \cdot c_3 \cdot c_4 > 0.$$

The single terms are each greater than zero for positive parameters.

It also holds that $x_w > 0$ for all A, c_1, c_2, c_3, $c_4 > 0$. The proof is given in Appendix B.2.

In the minimal mutated system, the following formula holds for the steady state of wild-type β-catenin (x_m):

$$x_m = \frac{1}{c_1}(c_2 - c_3 - c_4 \cdot A - c_5) + \frac{1}{c_1} \cdot \sqrt{(c_2 + c_3)^2 + (c_4 \cdot A)^2 + 2 \cdot c_4 \cdot A \cdot (-c_2 + c_3) + c_5 \cdot (2 \cdot c_2 + 2 \cdot c_3 + 2 \cdot c_4 \cdot A + c_5)} \quad (2.9)$$

with

$$c_5 = \frac{k_2 k_{-4} k_5 k_7}{k_6 k_{-7}}, \quad (2.10)$$

$c_5 > 0$.

The solution for x_m is real. The relation

$$(c_2 + c_3)^2 + (c_4 \cdot A)^2 + 2 \cdot c_4 \cdot A \cdot (-c_2 + c_3) + c_5 \cdot (2 \cdot c_2 + 2 \cdot c_3 + 2 \cdot c_4 \cdot A + c_5) > 0 \quad (2.11)$$

is always fulfilled for A, c_2, c_3, c_4, $c_5 > 0$. The transition to the inequation $(c_2 - c_4 \cdot A)^2 + c_3^2 + 2 \cdot c_2 \cdot c_3 + 2 \cdot A \cdot c_3 \cdot c_4 + c_5 \cdot (2 \cdot c_2 + 2 \cdot c_3 + 2 \cdot c_4 \cdot A + c_5) > 0$

shows that all single terms are positive. Thus, the sum is greater than zero. Furthermore, it also holds that $x_m > 0$, proof shown in Appendix B.2.

The availability of APC in the presence of both mutated and wild-type β-catenin is taken into consideration by the constant c_5 (equation 2.10). The higher c_5 is, the lower the availability of APC for the activation of the APC-dependent degradation of wild-type β-catenin. The term depends on the production (k_5) and degradation (k_6) of mutated β-catenin and the ratio of the rate constants of the formation (k_7) and the dissociation (k_{-7}) of the APC/β-catenin^{mut} complex. The more mutated β-catenin is produced (k_5) or the more mutated β-catenin binds APC (k_7), the higher c_5 is. Hence, a lower amount of APC is available. If more mutated β-catenin is degraded (k_6) or released from the APC/β-catenin^{mut} complex (k_{-7}), more APC is available to induce the degradation of wild-type β-catenin. In addition, the alternative degradation of wild-type β-catenin (k_2) and the release of β-catenin from the APC/β-catenin complex (k_{-4}) contribute to c_5. The higher these two parameters are, the less APC is bound to wild-type β-catenin and activates the APC-dependent degradation of β-catenin (reaction 3).

One can show that $x_m > x_w$ holds for all c_1, c_2, c_3, c_4, c_5, $A > 0$ (see Appendix B). Thus, the steady state of wild-type β-catenin in the mutated minimal model is always higher than in the minimal wild-type model. For $c_5 = 0$, the steady states of wild-type β-catenin in the wild-type and the mutated minimal model are equal. $c_5 \rightarrow 0$, if $k_6 \rightarrow \infty$ (degradation rate of mutated β-catenin) or $k_{-7} \rightarrow \infty$ (dissociation rate of the complex APC/β-catenin^{mut}). The same holds for $k_5 \rightarrow 0$ (production rate of mutated β-catenin) or $k_7 \rightarrow 0$ (binding rate of APC and β-catenin^{mut}). In these cases, the sequestration of APC by mutated β-catenin is prevented. On the one hand, this could be enabled by a fast turnover of mutated β-catenin. On the other hand, the sequestration of APC is abolished if APC cannot bind mutated β-catenin or if the APC/β-catenin^{mut} complex dissociates immediately. These conditions are related to mutated β-catenin. $c_5 \rightarrow 0$ also holds for $k_2 \rightarrow 0$ (alternative degradation rate of wild-type β-catenin) and $k_{-4} \rightarrow 0$ (dissociation rate of the complex APC/β-catenin^{wt}). These two reactions are related to wild-type β-catenin. For $k_2 \rightarrow 0$ it holds that $x_m \rightarrow \infty$. If $k_{-4} \rightarrow 0$, the impact of APC disappears, and consequently, the steady state of wild-type β-catenin only depends on the turnover of wild-type β-catenin. APC is the component that links wild-type and mutated β-catenin. In the case of $A = 0$ the steady state of β-catenin is equal in both the minimal wild-type and the minimal mutated model. The steady state of wild-type β-catenin is determined by its production and alternative degradation and calculated as $x_m = k_1/k_2$.

We next ask the question how the binding of APC and wild-type or mutated β-catenin (reactions 4 and 7, respectively) affect the unstimulated steady state of wild-type β-catenin. The relation of the binding constants and the steady state level is presented in Figure 2.26.

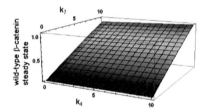

Figure 2.26: Dependence of wild-type β-catenin on the binding of the two β-catenin forms to APC.

The plot illustrates the dependence of the unstimulated steady state of wild-type β-catenin on the binding constants of APC to wild-type (k_4) and mutated β-catenin (k_7).

Figure 2.26 shows that the higher the rate constants are, the higher the unstimulated steady state of wild-type β-catenin. A change in the binding of mutated β-catenin to APC (k_7) has a stronger effect on the steady state level than an equal change in the binding of wild-type β-catenin and APC (k_4). An altered binding of APC and mutated β-catenin affects exclusively the level of APC that is available for the degradation of wild-type β-catenin. If the binding of wild-type β-catenin to APC is increased, this mediates two effects: i) more wild-type β-catenin is sequestered in the APC/β-catenin complex, but ii) the level of unbound APC is decreased that is available for the APC-dependent degradation of wild-type β-catenin (reaction 3). Therefore, the effect mediated by a change in the binding of wild-type β-catenin to APC is weaker than the effect mediated by a changed binding of APC and mutated β-catenin. For one particular parameter set, this is also shown by the sensitivity analysis (Figure 2.24). In a prospective and more general investigation of the impact of the individual processes on the unstimulated steady state of wild-type β-catenin, the sensitivity coefficients will be calculated analytically.

2.3.6   Discussion

To analyse the effects of a mutated form of β-catenin on the properties of the signalling pathway, a mathematical model of the Wnt/β-catenin pathway is developed that contains both wild-type and mutated β-catenin. The investigated β-catenin mutation prevents β-catenin to be phosphorylated and subsequently degraded.

The model can be used to analyse the effects if mutated β-catenin and wild-type β-catenin coexist in a system. Experimentally, both protein forms occur together in one cell if the cell carries one mutant and one wild-type allele of β-catenin. Mutated and wild-type β-catenin are also available at the same time in one cell if a vector containing the gene of mutated β-catenin is transfected into a wild-type cell. This case of transfection is regarded in the theoretical investigations. The general observations drawn from the investigations of the transfected case also hold for the case of mutation. The absolute concentration levels differ (analyses not shown).

The mathematical model including both β-catenin forms is developed and analysed. It is investigated how the presence of mutated β-catenin affects the steady state and the temporal behaviour of wild-type β-catenin. Theoretical predictions are drawn and partially verified by experiments performed in Rolf Gebhardt's group at University Leipzig. With regard to prospective investigations, a minimal model of the pathway is developed.

Reflection of published experimental observations

Using the developed detailed mathematical model one is able to reproduce published experimental observations. HepG2 cells are characterised by deletion of exon-3 of β-catenin in one allele. Hence, in these cells wild-type and mutated β-catenin coexist. The level of mutated β-catenin has been found to be higher than the level of wild-type β-catenin. The sum of mutated and wild-type β-catenin has been higher than the β-catenin level in wild-type cells

(Carruba et al., 1999). This is in agreement with our calculations and experiments. In other experiments using these cells it has been reported that the transcriptional activity has been increased (Satoh et al., 2000). The enhanced transcriptional activity has been experimentally decreased by transfection with APC or Axin (Satoh et al., 2000). Our simulations mimicking an overexpression of APC reveal that the wild-type β-catenin level is decreased compared to the scenario without APC overexpression (Figure 2.20). Proceeding on the assumption that the transcriptional activity reflects the β-catenin level, this finding corresponds to the experimental observations. The sensitivity analysis shows that an enhanced Axin production, that is comparable with an Axin transfection, leads to a decline in the β-catenin steady state level (Figure 2.18). Hence, this theoretical result also corresponds with the experimental observations.

Experiments using cells carrying a deletion in Ser[45] (HCT116 cells; human colon carcinoma cells) have been concentrated on the effects of the alternative β-catenin degradation. In these cells, β-catenin cannot be phosphorylated as the priming-phosphorylation cannot occur. If the alternative β-catenin degradation has been activated, a decrease of the β-catenin concentration, reporter gene activity as well as reporter gene expression have been measured (Gwak et al., 2009). This is in line with the simulation results. The sensitivity analysis reveals that an activation of the alternative β-catenin degradation results in a lower steady state of wild-type β-catenin (Figure 2.18).

Two effects of mutated β-catenin that might contribute to tumourigenesis

In comparison to the wild-type situation, in a transfected cell two important differences appear. They might contribute to tumourigenesis: i) in addition to wild-type β-catenin, mutated β-catenin exists that may regulate the expression of other target genes than wild-type β-catenin. ii) Due to the presence of mutated β-catenin, the concentration of wild-type β-catenin is increased compared to the wild-type cell.

Experimental observations have indicated that mutated β-catenin is able to regulate the expression of genes that are in part distinct from the genes regulated by wild-type β-catenin (Stahl et al., 2005; Zucman-Rossi et al., 2007). Gene expression profiles have been detected in experiments with wild-type liver cells and mutated liver cells. Mutated cells are either cells carrying a mutation in β-catenin or a mutation in another pathway components. The detected gene expression profiles have been differed depending on the presence and the kind of mutation (Stahl et al., 2005; Zucman-Rossi et al., 2007). Several reasons are conceivable, leading to these differences. They may be caused by the recognition of distinct binding motifs by mutated and wild-type β-catenin. Also the regulation or interaction with different transcriptional co-factors is reasonable. It is known that various co-factors are involved in β-catenin-mediated transcriptional regulation (Cadigan and Peifer, 2009; Hecht and Kemler, 2000; MacDonald et al., 2009).

The second effect of mutated β-catenin is to cause an increased steady state level of wild-type β-catenin in comparison to the wild-type situation (Figure 2.13 and Figure 2.14). Due to the presence of the mutated protein and its competition with wild-type β-catenin for binding partners, especially APC, the concentration of wild-type β-catenin increases in comparison to the wild-type model. Mutated β-catenin binds APC, leading to the reduction of APC that is available for the formation of the destruction complex. Consequently, less β-catenin is degraded via the destruction cycle. With regard of APC mutations, this effect is discussed in section 2.2; with regard to an overexpression of β-catenin, the effect is discussed by Goentoro and colleague (Goentoro and Kirschner, 2009). The resulting increased level of wild-type β-catenin may result in the activation of the target gene expression which in the wild-type cell is only induced upon pathway stimulation. The theoretical prediction of an increased concentration of wild-type β-catenin in the presence of mutated β-catenin has been verified experimentally (Figure 2.14).

It might be possible to reduce some effects mediated by mutated β-catenin by reducing the steady state concentration of wild-type β-catenin. Based on the sensitivity analysis we predict useful strategies of interventions. Besides the regulation of the turnover of mutated β-catenin, the regulation of the binding of APC and the two β-catenin forms is of high importance. The inhibition of the binding of APC and the two β-catenin forms would lead to a decrease of the concentration of wild-type β-catenin as less mutated β-catenin sequesters APC. Consequently, more APC is able contribute to the destruction complex, facilitating the degradation of wild-type β-catenin via the destruction cycle. Moreover, the advantage of the inhibition of these APC-specific reactions is that the reactions have a small impact on the wild-type β-catenin steady state in wild-type model but a strong impact in the model containing both β-catenin species. This means that such an inhibitor would affect tumour cells strongly but healthy cells marginally. For drug development this is an important feature as side effects on healthy cells should be as small as possible. An alternative to reduce the concentration of wild-type β-catenin in presence of the mutated protein would be an overexpression of APC. This has also been observed experimentally (Satoh et al., 2000). However, such an overexpression would also have strong effects in healthy cells. Therefore, the inhibition of the binding of APC to the two β-catenin forms is preferable, but it could be difficult to find a drug that inhibits the binding of APC and β-catenin and that does not affect the binding of β-catenin to the destruction complex.

Concerning the fold-change of wild-type β-catenin, the presence of mutated β-catenin causes a decrease of this characteristic (Appendix B.3). This is similar to the effect mediated by an APC mutation (discussed in section 2.2). The sensitivity analysis reveals that the perturbation of the APC/β-catenin complex formation (reaction 17), that is a potent drug target if the steady state is considered, has only a small effect on the steady state. To increase the fold-change of wild-type β-catenin strongly in the mutated model but only slightly in the wild-type model, appropriate perturbations of destruction reactions (reactions 4 to 9, 14 and 15) would be effective. Hence, distinct intervention strategies are necessary dependent on the specific property that should be affected. This is also valid in the case of APC mutations and already discussed in section 2.2.

Independent of whether the unstimulated β-catenin steady state or the β-catenin fold-change is considered, one has to keep in mind that with this approach of perturbations, only the mutation-mediated effects of the increased wild-type β-catenin level or the decreased fold-change, respectively, are affected. The effects of the gene expression regulated by mutated β-catenin are not attenuated.

2.4 Analyses of the effects of cellular E-cadherin levels and the Axin-2 feedback on the dynamics of β-catenin signalling

Different cell-types may respond differently to the same stimulus. Often, the underlying mechanism of this variability is unknown. One reason for the differences might be that regulatory features vary in distinct cell-types. For instance, concentrations of regulatory proteins differ. Recently, differences in the response of HEK293 (human embryonic kidney cells) and RK13 cells (rabbit kidney epithelial cells) after stimulation with an EGF (epidermal growth factor) stimulus have been analysed in a combined biochemical and theoretical approach (Kiel and Serrano, 2009). Different temporal behaviour and distinct responses to mutations have been explained by different feedback strengths.

We are interested in the dynamical behaviour of β-catenin in different cell-types. Upon pathway activation, time course data of β-catenin differ in murine primary hepatocytes and murine neuronal progenitor cells (C17.2 cells) (section 2.4.1). Furthermore, the concentrations of other pathway components differ in these two cell-types. On the one hand, the attention is directed to the effects of APC which is a known important regulator of Wnt/β-catenin signalling. On the other hand, experiments reveal differences in the E-cadherin and Axin-2 mRNA levels. These latter ones are related to important features that exist in cellular systems but not in extracts: i) E-cadherin binds β-catenin and is involved in cell-cell adhesion and ii) Axin-2 provides a transcriptional feedback loop. There is some evidence in the literature, that E-cadherin influences the dynamics of Wnt/β-catenin signalling in specific cell-types or under specific circumstances, but also that the β-catenin-dependent gene expression is unaffected by E-cadherin (Daugherty and Gottardi, 2007; Kuphal and Behrens, 2006; van de Wetering et al., 2001). The expression of Axin-2 is regulated by Wnt/β-catenin signalling. Axin-2 itself acts as component of this pathway. It participates in the destruction complex, leading to an increased degradation of β-catenin. Hence, Axin-2 mediates a negative feedback on β-catenin (Jho et al., 2002; Lustig et al., 2002). To understand the effects of E-cadherin and the Axin-2 feedback loop, the core-model of Wnt/β-catenin signalling is extended including these two cellular features. This allows a comparison of the core-model, which reflects conditions in an extract, and the model including E-cadherin and Axin-2, that represents a cellular scenario. What is the impact of the inclusion of E-cadherin and what effects are mediated by the Axin-2 feedback loop? How do these two extensions influence each other? How does APC affect Wnt/β-catenin signalling in combination with E-cadherin and the Axin-2 feedback loop? Are the distinct concentrations of pathway proteins responsible for the generation of the differences in the β-catenin dynamics that have been observed in the experiments? Thus, is it possible to describe the β-catenin dynamics in the two cell-types by the same mathematical model assuming different concentrations of pathway components? To address these questions we use a combined experimental and theoretical approach. All shown experiments have been performed by Frank Götschel, Monika Schempp or Katja Bruser in Andreas Hecht's group at University Freiburg.

2.4.1   Experimental data

β-catenin protein and Axin-2 mRNA time courses in hepatocytes and C17.2 cells

Time course data of β-catenin have been measured in murine primary hepatocytes as well as in murine neural progenitor cells (C17.2 cells) upon pathway activation by the application of SB216763, a chemical inhibitor of the GSK3. It reduces the capability of GSK3 to phosphorylate its targets. For each cell-type, the temporal behaviour of the β-catenin protein concentration and the concentration change of the mRNA level of the β-catenin target gene Axin-2 are shown in Figure 2.27, see also (Götschel, 2008; Götschel et al., 2008).

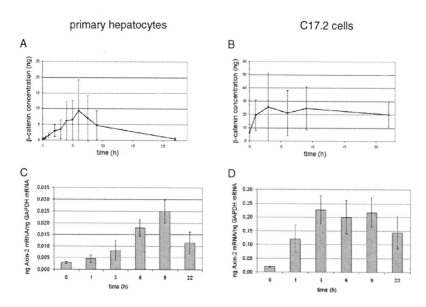

Figure 2.27: Time course data of β-catenin protein and mRNA of Axin-2 in primary hepatocytes and C17.2 cells.

The temporal behaviour of the absolute concentration of unbound β-catenin in (A) primary hepatocytes and (B) C17.2 cells is presented. The β-catenin concentration has been measured in ng per mg whole cell protein. The Axin-2 mRNA dynamics in (C) hepatocytes and (D) C17.2 cells are shown. The mRNA level of Axin-2 has been measured in ng per ng mRNA of the housekeeping gene GAPDH (Glyceraldehyde 3-phosphate dehydrogenase). The mean value with the respective standard deviation of three independent experiments is plotted.

Primary hepatocytes and C17.2 cells have been treated with 80 μM SB216763 and 25 μM SB216763, respectively. The stimuli are given at $t = 0$ h.

Due to pathway stimulation, the β-catenin concentration increases in both cell-types. The measured time courses of unbound β-catenin in the two cell-types differ in multiple aspects. Without stimulation (at $t = 0$ h), the absolute concentration of unbound β-catenin is lower in primary hepatocytes than in C17.2 cells. The concentration of β-catenin in hepatocytes (mean of 0.4 ng/mg whole cell protein) is around 15-fold lower than in C17.2 cell (mean of 6.1 ng/mg whole cell protein). While in primary hepatocytes the β-catenin concentration increases slowly reaching the maximal concentration around 6 h, the concentration raises much faster in C17.2 cells. In the latter cell-type, the maximum is reached around 3 h. In C17.2 cells, the β-catenin level remains high within the regarded time frame, even after 22 h.

In contrast, in primary hepatocytes, the concentration of β-catenin decreases after the maximum at 6 h. At 22 h, the β-catenin concentration has almost reached the same level as in the unstimulated case at $t = 0$ h. Under stimulated conditions, the absolute β-catenin concentration is lower in primary hepatocytes than in C17.2 cells. At the maximum, the β-catenin concentration in hepatocytes (mean of 9.3 ng/mg whole cell protein) reaches approximately the unstimulated steady state concentration of β-catenin in C17.2 cells. However, the relative concentration change in primary hepatocytes is higher than in C17.2 cells. In hepatocytes, the maximal relative concentration change is approximately 24 while the maximal relative concentration change in C17.2 cells is about five.

To summarise, in the two cell-types the absolute concentration levels of β-catenin differ as well as the relative concentration changes. In addition, the timing of the increase of the concentration and in the decreasing phase are different.

Additionally, the mRNA level of the β-catenin target gene Axin-2 has been measured in both cell-types. Under unstimulated conditions, the level of the Axin-2 mRNA is approximately 6.5-fold lower in primary hepatocytes (mean of 0.003 ng/ng GAPDH mRNA) than in C17.2 cells (mean of 0.02 ng/ng GAPDH mRNA). In primary hepatocytes as well as in C17.2 cells, the Axin-2 mRNA concentration follows the temporal behaviour of β-catenin. Its increase in C17.2 cells is much faster than in primary hepatocytes. In primary hepatocytes, the maximal concentration is not reached before 9 h. Therefore, there is a delay of at least 3 h between the maximum of β-catenin protein and the maximum of the Axin-2 mRNA. After 22 h, the β-catenin concentration reached almost the unstimulated steady state level in primary hepatocytes, whereas the Axin-2 mRNA level is reduced compared to the value at 9 h (factor of ~2.2) but still elevated compared to the unstimulated Axin-2 mRNA concentration (factor of ~3.6). In C17.2 cells, a delay between the dynamics of Axin-2 mRNA and the β-catenin dynamics has not been detected.

A summary of the dynamical features experimentally observed in the two cell-types is provided in Table 1.

Concentrations of selected pathway components in primary hepatocytes and C17.2 cells

For the further characterisation of the two cell-types, the concentrations of other pathway components have been analysed. These are APC and E-cadherin. Regarding E-cadherin, there are large differences between primary hepatocytes and C17.2 cells both on the level of protein and on the level of mRNA concentration (see Figure 2.28B). The concentration of the E-cadherin mRNA is 1000-fold higher in hepatocytes than in C17.2 cells (upper part of Figure 2.28B). Also the protein level is much higher in primary hepatocytes than in C17.2 cells, shown qualitatively by the Western blot in the upper part of Figure 2.28A. In contrast, the mRNA concentration of APC is higher in C17.2 cells than in primary hepatocytes (lower part of Figure 2.28B). However, the difference is much smaller compared to the difference in the mRNA level of E-cadherin. In hepatocytes, the APC mRNA is expressed 1.5–2-fold less than in C17.2 cells. Unfortunately, it has not been possible to detect the protein level of APC in primary hepatocytes. Hence, a comparison of the APC protein levels in the two cell-types is not possible. The properties are summarised in Table 1.

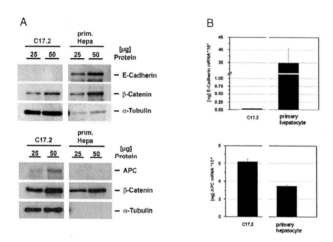

Figure 2.28: Protein and gene expression of APC and E-cadherin in primary hepatocytes and C17.2 cells.

(A) The protein expression is analysed by Western blots. The upper part focuses on E-cadherin while the lower part aims attention on APC. The APC protein level in hepatocytes ("prim. Hepa") is undetectable by Western blot. The gels are loaded with 25 μg or 50μg whole cell protein (indicated by 25 or 50, respectively). (B) The amount of gene expression is analysed by quantitative real time PCR. The upper diagram shows the data of the E-cadherin mRNA; the lower diagram shows the data of the APC mRNA. The mean value and standard error of three independent experiments are shown.

Table 1: Cell-type specific properties.

Concentration of E-cadherin, APC, Axin-2 mRNA and β-catenin are summarised for the two investigated cell-types. The expression of Axin-2 is related to the expression of the housekeeping gene GAPDH. The mean values and standard deviations are given. The temporal behaviour of β-catenin and Axin-2 mRNA are described.

	primary hepatocyte	C17.2 cell
E-cadherin mRNA	mean of 0.0035 ng	mean of ~0.00003 ng
E-cadherin protein	very high	low
APC mRNA	mean of 0.00035 ng	mean of 0.00061 ng
APC protein	unknown, low concentration assumed	high
β-catenin steady state (unstimulated)	0.4 ± 0.4 ng/mg whole cell protein	6.1 ± 6.3 ng/mg whole cell protein
β-catenin maximum	9.3 ± 9.9 ng/mg whole cell protein at 6 h after stimulation	25.6 ± 25.6 ng/mg whole cell protein at 3 h after stimulation
temporal behaviour of β-catenin	A slow increase (till 6 h) is followed by a considerable decrease. At 22 h, the unstimulated β-catenin level is almost reached again.	The β-catenin concentration increases rapidly reaching the maximum at 3 h. Between 3 h and 22 h, almost no decline is observed.
unstimulated level of Axin-2 mRNA	0.003 ng/ng GAPDH mRNA	0.02 ng/ng GAPDH mRNA
maximal level of Axin-2 mRNA	0.025 ± 0.005 ng/ng GAPDH mRNA, 9 h after stimulation	0.23 ± 0.05 ng/ng GAPDH mRNA at 3 h after stimulation
temporal behaviour of Axin-2 mRNA	The temporal behaviour of Axin-2 mRNA follows the time course of β-catenin (increase, decrease). There is a time delay between the two time courses: The maximal Axin-2 mRNA concentration is achieved at least 3 h later.	The temporal behaviour of Axin-2 mRNA follows the time course of β-catenin. There is no delay detectable between β-catenin dynamics and dynamics of Axin-2 mRNA.

2.4.2   Model including E-cadherin and the feedback loop via Axin-2

We address the question whether the different temporal behaviour of β-catenin that has been observed experimentally can be explained by the same underlying mechanism. Therefore, we extend the core-model by inclusion of i) the binding of β-catenin and E-cadherin as well as ii) the feedback loop via Axin-2. These are two properties that occur in cells but not in extracts. In this context, the core-model is referred to as extract model while the extended model is referred to as cellular model. The reactions that are the same in the extract model and the cellular model (reactions 1 to –17) are referred to as core-reactions in the following. A scheme of the cellular model is shown in Figure 2.29.

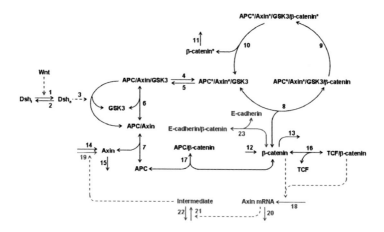

Figure 2.29: Scheme of the cellular model.

The reactions of the extract model are shown in black (processes 1 to 17). Model extensions towards E-cadherin binding of β-catenin are coloured in green while reactions of the Axin-2 feedback loop are coloured in red. Components in a complex are separated by a slash. The phosphorylation of a component is indicated by an asterisk. Numbers next to arrows denote the number of the particular reaction. One-headed arrows denote reactions taking place in the indicated direction. Dashed arrows represent activation steps and double-headed arrows illustrate binding reactions.

The extract model is extended by the inclusion of the reversible binding of β-catenin and E-cadherin (reactions 23 and –23). These reactions and components are coloured in green in the model scheme shown in Figure 2.29. It is assumed that the total concentration of E-cadherin obeys a conservation relation. Thereby we follow the assumption made by Lee and colleagues (Lee et al., 2003) for APC, GSK3, Dsh and TCF that the concentration remains constant within the regarded time frame.

Red-coloured reactions and components participate in the Axin-2 feedback loop. TCF/β-catenin and β-catenin activate the production of the mRNA of Axin-2 (reaction 18). The mRNA activates the production of an intermediate (reaction 21) which could represent a modified mRNA. The intermediate is able to activate the synthesis of Axin-2 (reaction 19). Axin-1 and Axin-2 build the Axin pool. In the model, it is not distinguished between the two species. The mRNA and the intermediate are degraded via reactions 20 and 22, respectively. In part, we follow the realisation of the Axin-2 feedback loop introduced by Cho and colleagues (Cho et al., 2006). They have supposed that TCF/β-catenin as well as unbound β-catenin are able to directly activate the expression of Axin-2. Due to this straight connection, the delay of the effect of the feedback loop on the β-catenin concentration is neglected. For their analyses on the β-catenin steady state, this model has been valuable. As we are also interested in the effect of the feedback loop on the dynamical properties of the model, their implementation of the feedback loop is not appropriate. Hence, we follow the assumption of Cho and colleagues to consider TCF/β-catenin and unbound β-catenin as activators of the gene expression, but assume that these species induce the production of the Axin-2 mRNA. We adopt the assumption of Cho and colleagues to neglect further regulators of the gene expression.

Please, note that there is yet no compartmentation in cytoplasm and nucleus included. Experimental data reveals that there is almost no delay between the β-catenin dynamics in the cytoplasm and in the nucleus ((Götschel, 2008), data not shown). Hence in this context, a separate consideration of these two compartments is not necessary.

The model is given by equations A.1 to A.10 and A.13 to A.37, listed in Appendix A, and C.1 to C.13, listed in Appendix C. Our analyses focus on the effects of different total APC and total E-cadherin levels as well as different strengths of the feedback loop. Therefore, only these parameters are varied while all other parameters are kept as in the extract model. The feedback loop strength is altered by variation the degradation rate of the intermediate (k_{22}). Thereby, the feedback-loop-dependent Axin production is regulated. The higher the degradation rate, the weaker the feedback loop is, as less Axin is produced via this mechanism.

The rate constants and concentrations related to the core-reactions are given in Table 5, Appendix A. The additional concentrations and rate constants of the cellular model are provided in Table 11, Appendix C.

2.4.3   Effects of the single model extensions

Effects of the binding of β-catenin and E-cadherin

First, the effects of the inclusion of E-cadherin are investigated. Thus, only the black and green reactions and components (scheme in Figure 2.29) are taken into account. In Figure 2.30, the effects of different total concentrations of E-cadherin on the temporal behaviour of β-catenin are shown.

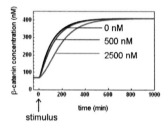

Figure 2.30: Effects of the inclusion of E-cadherin on the β-catenin dynamics.
The constant stimulus is applied at $t = 0$ min. The total concentration of E-cadherin is written at the particular curve (0 nM, 500 nM, 2500 nM). The black line (0 nM) shows the β-catenin dynamics of the extract model.

Figure 2.30 illustrates that the steady state concentrations of β-catenin are not affected by the inclusion of the binding of β-catenin and E-cadherin. In contrast, the temporal behaviour of β-catenin changes if E-cadherin is present. The more E-cadherin is available, the more decelerated the β-catenin increase is upon pathway stimulation. To investigate the deceleration, the time point t_Δ is regarded at which 50% of the concentration difference between stimulated and unstimulated steady state are reached. For *E-cadherin* = 0 nM, $t_\Delta = 84.5$ min. The higher the total concentration of E-cadherin is, the higher the value of t_Δ. If *E-cadherin* = 500 nM, $t_\Delta = 100.6$ min and if *E-cadherin* = 2500 nM, $t_\Delta = 165.5$ min. E-cadherin buffers the increase of the β-catenin concentration.

Effects of the feedback loop via Axin-2

Next, we are interested in the effect of the exclusive inclusion of the feedback loop via Axin-2. The black and red reactions of the model shown in Figure 2.29 are considered. The β-catenin dynamics upon pathway stimulation is shown in Figure 2.31. The strength of the feedback is altered by the variation of the degradation rate of the intermediate (k_{22}).

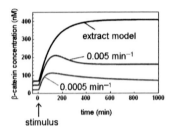

Figure 2.31: Effect of the feedback loop via Axin-2 on the β-catenin dynamics.
The constant stimulus is applied at $t = 0$ min. The black line shows the β-catenin dynamics of the extract model. The red lines show the temporal behaviour of β-catenin if the Axin-2 feedback is included. Two different strengths of the feedback are assumed. The respective k_{22}-values (0.005 min^{-1}, 0.0005 min^{-1}) are written at the particular curve.

The inclusion of the feedback loop causes changes in the unstimulated and stimulated steady state levels of β-catenin as well as the temporal behaviour (Figure 2.31). The stronger the feedback is, the lower the unstimulated steady state concentration of β-catenin. This is caused by the fact that even without pathway stimulation the feedback loop is active resulting in a higher Axin concentration. Consequently, a higher amount of the destruction complex is built and a higher amount of β-catenin is degraded via the destruction cycle.

After administration of a stimulus, the absolute β-catenin concentration increases. The stronger the feedback, the lower the β-catenin maximum is. Furthermore, the feedback loop enables the appearance of the decreasing phase of the β-catenin concentration. It causes an overshot of the β-catenin concentration. The stimulated β-catenin steady state is no longer equal to the β-catenin maximum. The higher the β-catenin concentration increases, the stronger the feedback loop is activated. The concentration of β-catenin begins to decrease until it reaches its stimulated steady state level.

2.4.4   Effects of E-cadherin, APC and Axin-2 feedback

In this section, the effects of the combined extensions are investigated. We are interested in the effects on the steady state and the temporal behaviour of β-catenin. Concerning the temporal behaviour, we are interested in the effects on the increasing and decreasing phase, mainly focussing on the increasing phase. The period between the time point of the stimulus administration and the time point the maximum is reached is referred to as increasing phase. The decreasing phase is the time span between the maximum and the reaching of the stimulated steady state. In the cases in which no maximum occurs, the increasing phase is the time interval between the time point of stimulation and the time point at which 99.5% of the stimulated steady state level is reached.

The experiments have shown that the concentrations of pathway components vary between the cell-types (section 2.4.1). We focus on the effects of three parameters: i) the total E-cadherin concentration, ii) the total APC concentration, and iii) the strength of the feedback via Axin-2. Both E-cadherin and APC obey conservation relations in the model, implying that their total concentrations do not change over time. A change in the APC or E-cadherin concentration always means a change in the total concentration of the respective component.

The effects of these three parameters on different signalling characteristics are investigated. These are the unstimulated steady state concentration of β-catenin, the maximal β-catenin concentration, the ratio between these concentrations and the timing of the increasing phase. The understanding of the impact of the parameters on these characteristics will contribute to the understanding of the cell-type-specific experimental observations.

The impact of all model parameters on the unstimulated steady state and the β-catenin dynamics is investigated by a sensitivity analysis in section 2.4.6.

Impact of APC and feedback strength on the unstimulated steady state of β-catenin

The unstimulated steady state concentration of β-catenin does not only depend on the feedback strength (see Figure 2.31, section 2.4.3) but is also strongly dependent on the APC concentration. The experiments have shown different APC levels in the two analysed cell-types (section 2.4.1). It is shown theoretically that the steady state of β-catenin is independent of the total concentration of E-cadherin (see Figure 2.30, section 2.4.3).

The dependence of the unstimulated steady state level of β-catenin on both the total APC concentration and the feedback strength is shown in Figure 2.32.

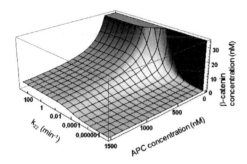

Figure 2.32: Dependence of the unstimulated β-catenin steady state on the APC concentration and the feedback strength.

The feedback strength is altered by variation of the degradation rate of the intermediate (k_{22}). The k_{22}-values are plotted on a logarithmic scale. The β-catenin concentration is cut off at 35 nM. For total $APC = 0$ nM, β-catenin reaches its maximal concentration of ~1646 nM (not shown).

Figure 2.32 shows that the unstimulated β-catenin steady state level decreases with an increasing APC concentration. The more APC is available, the more β-catenin binds APC and the more β-catenin is degraded via the destruction cycle as more destruction complex is built.

The unstimulated steady state concentration of β-catenin also decreases with increasing feedback strength. In this case, a higher concentration of the destruction complex arises from a higher concentration of Axin whose production is increased by the feedback (see Figure C.1, Appendix C, for corresponding total Axin levels).

For total $APC = 0$ nM, the concentration of β-catenin is independent of the strength of the feedback loop. This is due to the fact that in the absence of APC, no destruction complex is formed. Therefore, the β-catenin concentration is only determined by the production and alternative degradation, and thus, independent of the concentration of Axin and the strength of the feedback loop. The maximal β-catenin level of ~1646 nM is reached and the inducibility of a β-catenin response caused by stimulation is lost.

Impact of APC, E-cadherin and the feedback strength on the β-catenin maximum

The impact of the total E-cadherin concentration and the total APC concentration on the β-catenin maximum is investigated for three exemplary feedback strengths. The β-catenin maximum is plotted colourcoded in Figure 2.33.

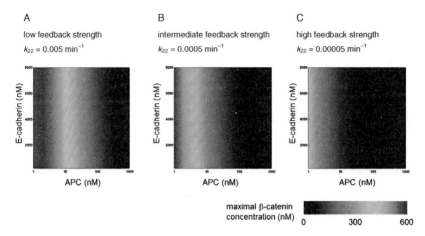

Figure 2.33: Impact of the total APC level, the total E-cadherin level and the strength of the feedback on the β-catenin maximum.
For (A) $k_{22} = 0.005$ min^{-1}, (B) $k_{22} = 0.0005$ min^{-1}, and (C) $k_{22} = 0.00005$ min^{-1}, the impact of the total APC concentration and the total E-cadherin concentration on the β-catenin maximum is investigated. The β-catenin maximum is plotted colourcoded for the respective concentrations (see legend). The APC concentration is plotted on a logarithmic scale.

The analysis presented in Figure 2.33 confirms that the β-catenin maximum strongly depends on the strength of the feedback loop (see also Figure 2.31). It holds that the stronger the feedback loop is, the lower the β-catenin maximum. If the feedback is strong, more Axin is produced, leading to a higher concentration of the destruction complex. Hence, more β-catenin is bound to the destruction complex and subsequently degraded. Its maximum declines. In addition, it is shown that the maximum is highly dependent on the APC concentration. The higher the APC concentration is, the lower the β-catenin maximum. As more APC is available in the system, more β-catenin is degraded in the APC-dependent manner. The concentration of unbound β-catenin decreases. The impact of the total E-cadherin concentration on the β-catenin maximum is very small.

Impact of APC, E-cadherin and the feedback strength on the extent of β-catenin increase

The ratio r of the β-catenin maximum and the unstimulated β-catenin steady state level is analysed with respect to the feedback strength, the total APC and the total E-cadherin concentrations

$$r = \frac{\beta\text{-catenin maximum}}{\text{unstimulated }\beta\text{-catenin steady state}}.$$ (2.12)

In the following this ratio is referred to as extent of increase.

In Figure 2.34, the ratio r is plotted colourcoded in the area spanned by the total E-cadherin and the total APC concentration for the three different feedback strengths also considered in the paragraph above.

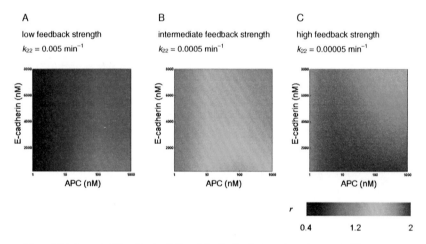

Figure 2.34: Impact of E-cadherin, APC and the feedback strength on the extent of increase. For (A) $k_{22} = 0.005$ min^{-1}, (B) $k_{22} = 0.0005$ min^{-1}, and (C) $k_{22} = 0.00005$ min^{-1}, the impact of the total APC concentration and the total E-cadherin concentration on the extent of β-catenin increase (ratio r) is examined. The total APC concentration is plotted on a logarithmic scale. The calculated r-values are plotted colourcoded for the respective concentrations (see legend).

Figure 2.34 reveals that the extent of the increase of β-catenin depends strongly on the strength of the feedback loop and only slightly on the total levels of E-cadherin and APC. The higher the feedback strength is, the greater r. However, the absolute levels of the β-catenin steady state and maximum are low compared to weaker feedback strengths (see Figure 2.32 and Figure 2.33). The ratio r also increases with increasing total APC concentrations. As seen before, high APC concentrations cause low β-catenin levels (Figure 2.32 and Figure 2.33). Again, a higher extent of increase is possible. Compared to the effect mediated by the feedback strength, this impact is low.

For high APC concentrations, the E-cadherin level slightly influences the extent of increase. The more E-cadherin exists in the system, the smaller the ratio r. If the APC concentration is high, the β-catenin concentration is low and E-cadherin strongly buffers β-catenin. Due to the feedback mechanism, more β-catenin is degraded, resulting in a slight decrease of the β-catenin maximum. The ratio r decreases. Thus, this decrease is related to a delayed increase of the β-catenin concentration and a more sustained activated feedback mechanism. However, in comparison to the impact of the feedback strength and the total APC concentration, the impact of the total E-cadherin level is very small.

Impact of APC, E-cadherin and the feedback strength on the increasing phase of β-catenin

Next, the impact on the timing of the increasing phase is examined. Due to a stimulus, the β-catenin concentration increases over time. Due to the feedback, it may decrease thereafter. Here, it is focussed on the increasing phase which is described by the time span between the time point of pathway stimulation and the time point at which the β-catenin maximum is reached (t_{max}).

t_{max} is plotted colourcoded in the area spanned by the two considered concentrations (Figure 2.35). A comparison of the impact on t_{max} and on other measures characterising the timing of the β-catenin maximum is presented in section 2.4.6. There, also the impact of the other model parameters on t_{max} is investigated.

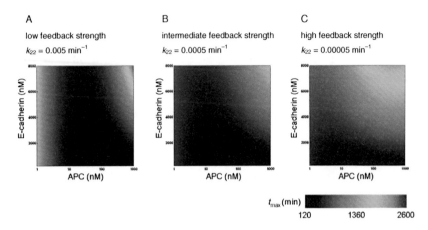

Figure 2.35: Effect of the APC concentration, the E-cadherin concentration and the feedback strength on the increasing phase.

For (A) $k_{22} = 0.005$ min^{-1}, (B) $k_{22} = 0.0005$ min^{-1}, and (C) $k_{22} = 0.00005$ min^{-1}, the impact of the total APC concentration and the total E-cadherin concentration on the increasing phase of β-catenin (t_{max}) is investigated. The t_{max}-values are plotted colourcoded for the respective concentrations (see legend). The APC concentration is plotted on a logarithmic scale.

Figure 2.35 shows that t_{max} is determined by all three parameters: i) the total concentration of APC, ii) the total concentration of E-cadherin, and iii) the feedback strength. First, we concentrate on the effects of APC concentrations higher than 10 nM. Effects of lower APC concentrations are discussed later in this paragraph.

In the case of APC concentrations larger than 10 nM, in general it holds that the t_{max}-value increases with increasing total APC and E-cadherin concentrations. E-cadherin buffers β-catenin. This leads to a delayed increase, as already seen in Figure 2.30. The more E-cadherin exists in the system, the more β-catenin can be bound and the longer the increasing phase lasts. The t_{max}-value increases. With an increasing concentration of APC, the concentration of β-catenin decreases. In the cases of lower β-catenin levels, the buffering effect of E-cadherin is stronger developed than in cases of higher β-catenin concentrations. Therefore, t_{max} is the highest for high total APC and high total E-cadherin concentrations.

The increasing phase is also strongly dependent on the strength of the feedback. It influences the increasing phase by two effects: i) It affects the β-catenin steady state concentration, and ii) it enables the decreasing phase. For the same total E-cadherin and APC levels, t_{max} changes with varying feedback strengths. With the exception of a low feedback strength and high total

APC and E-cadherin concentrations it holds that the higher the feedback strength is, the higher the t_{max}-value. The stronger the feedback loop is, the lower the β-catenin concentration. This leads to a decelerated increase of the β-catenin concentration since the buffering effect mediated by E-cadherin is stronger developed.

Next, we address the question why the time point of the maximum is shifted to higher values in the case of a weak feedback and high concentrations of APC and E-cadherin. Furthermore, we ask why the t_{max}-value is higher for very low APC concentrations than for intermediate APC concentrations. This effect is observed in the case of a weak feedback loop. Under both conditions – i) high total APC and E-cadherin levels as well as ii) very low APC concentrations – the observed effect is based on the same mechanism. In both cases, no decreasing phase of the β-catenin concentration occurs. It lasts longer to reach 99.5% of the stimulated steady state level. While for the intermediate feedback strength even for high E-cadherin levels a decreasing phase appears, a weak feedback loop is not able to induce a decreasing phase if the APC level and the E-cadherin level are very high. The t_{max}-value increases. For all analysed feedback strengths, a combination of APC and E-cadherin concentration exists at which a decreasing phase does not occur. The β-catenin concentration increases until it reaches its stimulated steady state level. The particular total APC and E-cadherin levels differ dependent on the strength of the feedback. The absence of the decreasing phase is also the reason for higher t_{max}-values in the case of low APC levels, especially seen in the case of a weak feedback loop in Figure 2.35. If the APC level is very low, the β-catenin concentration is very high and increases after stimulation on an even higher stimulated steady state level. This takes longer compared to the time until the maximum is reached if a decreasing phase is induced in the case of slightly higher APC concentrations.

In the analysis shown in Figure 2.35, the feedback strength varies between $k_{22} = 0.005$ min^{-1} and $k_{22} = 0.00005$ min^{-1}. In Figure 2.36, the effects on the increasing phase are shown for one particular combination of total E-cadherin and APC level if the feedback strength ranges from $k_{22} = 1$ min^{-1} to $k_{22} = 10^{-7}$ min^{-1}.

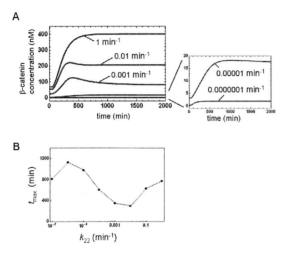

Figure 2.36: β-Catenin dynamics for different feedback strengths.

(A) The β-catenin dynamics upon constant pathway activations are presented, assuming different degradation rates of the intermediate (k_{22}). The respective values are written at the particular curves. The inset presents the temporal behaviour for very low k_{22}-values in a smaller concentration frame. (B) t_{max} is plotted against the k_{22}-value. The k_{22}-values are plotted on a logarithmic scale. If no decreasing phase occurs, t_{max} is calculated if 99.5% of the steady state concentration is reached.
For all calculations the following total concentrations are considered: total E-cadherin = 2500 nM and total APC = 100 nM. The pathway is stimulated at $t = 0$ min.

Depending on the degradation rate of the intermediate (k_{22}), the unstimulated and stimulated steady states of β-catenin vary as well as the temporal behaviour (Figure 2.36). The higher the degradation rate of the intermediate, the higher the unstimulated and the stimulated steady state levels are. If the degradation rate is high, less intermediate exists. Therefore, less Axin is produced in dependence of the intermediate. The strength of the feedback loop does not only affect the steady state but also the temporal behaviour of β-catenin.

In the extreme cases of very high and very low k_{22}-values, no decreasing phase is observed. Upon pathway stimulation, the β-catenin concentration increases and gains its stimulated steady state level without a period in which its concentration transiently decreases. For high degradation rates, the feedback loop is very weak and loses its capability to induce a decrease of the β-catenin concentration. If the feedback is very strong, there is also no decline visible. Due to the strong feedback loop, the β-catenin concentration is very low and less E-cadherin/β-catenin complex is built. Therefore, the concentration of unbound E-cadherin is very high. Under these circumstances, the stimulus-induced increase of β-catenin is immediately buffered by the binding to E-cadherin. Furthermore, the feedback loop is maximally activated and cannot be increased any further. The combination of both effects prevents the occurrence of an overshoot of the β-catenin concentration. In these cases of very weak and strong feedbacks, the β-catenin concentration increases reaching the stimulated steady state level. The time until 99.5% of the stimulated steady state level is reached differs. However, if the feedback is very weak ($k_{22} = 1$ min^{-1}) or strong ($k_{22} = 0.00001$ min^{-1}), this increasing phase of the β-catenin concentration lasts long. Intermediate feedback strengths provide the possibility of a decreasing β-catenin concentration following an initial increase. The feedback loop is strong enough to induce the decrease and weak enough to be inducible and to diminish the concentration of unbound E-cadherin in the unstimulated state. The increasing phase depends on the strength of the feedback and the resulting concentrations of β-catenin and unbound E-cadherin. The weaker the feedback is, the earlier the maximum is reached. Less unbound E-cadherin is available for buffering β-catenin (Figure 2.36B).

Summary of the effects of the total concentrations of APC and E-cadherin and the feedback strength

Taken together, the investigations in this section reveal different effects of the three analysed parameters on the different investigated signalling characteristics. The strength of the feedback loop has a strong effect on all analysed characteristics: the unstimulated steady state of β-catenin, the β-catenin maximum, their ratio as well as on the increasing phase. The stronger the feedback loop is, the lower the β-catenin steady state and the β-catenin maximum. Their ratio is increased. For intermediate total APC and E-cadherin concentrations, a stronger feedback loop causes a shorter increasing phase.

APC affects the steady state and the maximum of β-catenin strongly, but has only a weak effect on their ratio. It holds that the higher the total APC concentration, the lower the β-catenin steady state and the β-catenin maximum are. Higher concentrations of APC also cause a longer increasing phase.

E-cadherin has a strong effect on the increasing phase. The β-catenin steady states are independent of E-cadherin. The maximal β-catenin concentration and the extent of increase are almost not affected by changes in the total E-cadherin level.

The increasing phase is affected by all three parameters. This characteristic is further investigated by a sensitivity analysis in section 2.4.6.

The effects of the two model extensions – the inclusion of the feedback loop via Axin-2 and the binding of E-cadherin and β-catenin – interact. The stronger the feedback loop is, the lower the concentration of β-catenin. In this case, E-cadherin has a stronger effect on the increasing phase. In contrast, if the feedback loop is weak, the β-catenin steady state concentration is high and the buffering effect of the same concentration of E-cadherin is lower. These effects also interact with the effects mediated by APC. The higher the total APC concentration is, the lower the β-catenin steady states and the stronger the effect of the total level of E-cadherin on the increasing phase.

2.4.5   Combination of experiments and simulations

After the analyses of numerous properties of the cellular model, we examined whether we can semiquantitatively reproduce the temporal behaviour of the β-catenin protein and the mRNA of Axin-2 as measured in murine primary hepatocytes and neural progenitor cells (C17.2 cells) (Figure 2.27). We assume that the cellular model represents the network structure of Wnt/β-catenin signalling in both cell-types. The analyses in the previous section 2.4.4 reveal that different total APC and E-cadherin concentrations as well as a distinct strength of the feedback loop can generate the differences that have been observed experimentally in the two cell-types. Hence, with exception of these three parameters, the same concentrations and rate constants are used in the simulations for the representation of both cell-types. In the following, the cellular model in which the parameters are used that are assumed for the representation of C17.2 cells is referred to as C17.2-cell-like model; the cellular model taking into account the parameters assumed for representation of the primary hepatocytes is referred to as hepatocyte-like model.

Temporal behaviour of β-catenin and the mRNA of Axin-2 in the hepatocyte-like and C17.2-cell-like model

With different total APC and total E-cadherin concentrations and distinct strengths of the Axin-2 feedback the temporal behaviour of β-catenin in the two cell-types is simulated (Figure 2.37). The postulated differences of the parameters are motivated by the experimental observations (summarised in Table 1). Compared to the C17.2-cell-like model, we assume a higher total concentration of E-cadherin in the hepatocyte-like model, as seen in the experiments, a lower total APC concentration, as indicated by the lower mRNA level, and a stronger feedback. The detailed parameter values are provided in Table 10, Table 11 and Table 12 in Appendix C. These parameters are set based on the results of the analyses presented in section 2.4.4. They are not fitted as for that the experimental data is insufficient.

A   B

β-catenin dynamics   Axin-2 mRNA dynamics

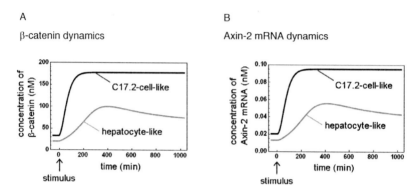

Figure 2.37: Simulated temporal behaviour of β-catenin and the mRNA of Axin-2 in the two cellular models.

In (A), the simulated β-catenin dynamics, and in (B) the dynamics of the Axin-2 mRNA are shown. The time courses simulated in the C17.2-cell-like system are coloured in blue, the hepatocyte-like time courses are presented in purple. For the representation of both cell-types the same mathematical model is used but partially different parameter sets. For the C17.2-cell-like time courses, a lower total level of E-cadherin (300 nM), a higher total level of APC (200 nM) and a weaker feedback ($k_{22} = 0.05$ min^{-1}) compared to the hepatocyte-like system is considered. The hepatocyte-like time courses are simulated assuming the following total concentrations and rate constant: $E\text{-}cadherin = 2500$ nM, $APC = 100$ nM and $k_{22} = 0.0005$ min^{-1} (see also Table 10, Table 11 and Table 12 in Appendix C). All other parameters are equal in both models and given in Table 5 and Table 7, Appendix A. The system is constantly stimulated at $t = 0$ min.

Figure 2.37 shows that in general, the qualitative temporal behaviour seen in experiments (shown in Figure 2.27) is semiquantitatively reproduced in the simulations. We focus on four characteristics of the β-catenin dynamics for which the experimental observations reveal that they differ in the two cell-types: i) the unstimulated steady state concentrations of β-catenin, ii) the β-catenin maxima, iii) the increasing, and iv) the decreasing phases (see section 2.4.1). The simulations show that in the C17.2-cell-like model, the unstimulated steady state concentration of β-catenin is about 1.6-fold higher than in the hepatocyte-like model. Upon pathway stimulation, the β-catenin concentration increases in both models. According to the experimental data, in the C17.2-cell-like model the β-catenin concentration rises faster than in the hepatocyte-like system. The β-catenin maximum is reached later in the hepatocyte-like model than 99.5% of the stimulated steady state level are reached in the C17.2-cell-like model. The increasing phase in the C17.2-cell-like and hepatocyte-like model lasts ~277 min and ~400 min, respectively. Furthermore, the β-catenin maximum is higher in the C17.2-cell-like model (~177 nM) than in the hepatocyte-like model (~100 nM). In the C17.2-cell-like model, the β-catenin concentration stays on a high level and does not decrease. This is in accordance with the experimental observations. In the hepatocyte-like system, a decreasing phase of the β-catenin concentration is simulated as seen in the experiments. Between the maximum at $t \sim 400$ min and $t = 1000$ min, the β-catenin concentration decreases from ~100 nM to ~75.3 nM. The stimulated steady state is not reached at $t = 1000$ min. The stimulated steady state concentration is ~57.5 nM, and hence, about 2.9-fold higher than the unstimulated steady state level. This is in contrast to the experiments in which the β-catenin concentration reaches almost the unstimulated steady state level after 22 h. The analysis presented in Figure 2.36 shows that one cannot circumvent this deviation between the simulations and the experimental observations by just increasing the feedback strength. A further deviation between experimental and simulated results is that the absolute β-catenin concentrations in the two cell-types differ experimentally to a higher extent than in the simulations. Also the relative concentration changes are stronger developed in the experimental data than in the simulations. The experimental data shows that the maximal β-catenin concentration in hepatocytes is almost equal to the unstimulated β-catenin level in C17.2 cells. The relative concentration change of β-catenin in primary hepatocytes is about 5-fold higher than the relative concentration change in C17.2 cells (see section 2.4.1). In the simulations, the maximal β-catenin concentration in the hepatocyte-like system (~100 nM) reaches almost the β-catenin concentration in the C17.2-cell-like system at which 50% of the concentration difference between stimulated and unstimulated steady state is reached (~104.5 nM). The relative concentration changes are very similar in both cellular models. In the C17.2-cell like model, the ratio between the stimulated and the unstimulated steady state is approximately 5.5. In the hepatocyte-like model, the ratio between the unstimulated β-catenin steady state and the β-catenin maximum is ~5. Except for these deviations, the model reproduces the qualitative β-catenin behaviour in the two cell-types very well.

Like the temporal behaviour of β-catenin, also the dynamics of the mRNA of Axin-2 are simulated and compared to the experimental findings. The steady state level in the hepatocyte-like model is 0.013 nM, and thus, about 1.5-fold lower than in the C17.2-cell-like model. In both models, the mRNA dynamics follow the temporal behaviour of β-catenin. These properties are in accordance with the experimental observations (section 2.4.1). However, the delay between the β-catenin dynamics and the Axin-2 mRNA dynamics detected in primary hepatocytes is not reproduced by the model. The maximum of the Axin-2 mRNA is reached at $t \sim 410$ min, thus, just 10 min delayed compared to the β-catenin maximum. To circumvent this deviation between simulation and experimental observation, an additional reaction upstream of the production of the Axin-2 mRNA but downstream of β-catenin and β-catenin/TCF would be reasonable. This reaction can generate a further delay. Experiments reveal that there is no delay between the β-catenin dynamics in the cytoplasm and the nucleus (Götschel, 2008). Hence, model extensions would focus on regulatory mechanisms of gene expression and not on the nucleo-cytoplasmic shuttle of β-catenin. Till today, less is known about the regulatory mechanisms of the gene expression. Therefore, they are not yet included.

Elucidation of the pathway deactivation in primary hepatocytes

As discussed in the previous paragraph, there is a deviation between experimental data and simulation in the decreasing phase of the β-catenin concentration in primary hepatocytes. This raises the question whether other mechanisms could be the reason for the experimentally observed decline of the concentration. One hypothesis concentrates on the fate of the substance mediating the pathway activation. The experiments and also the model simulations are performed with a constant stimulus. In the experiments, an artificial ligand – SB216763 – has been used for pathway activation via GSK3 inhibition. Hepatocytes are involved in processes of detoxification. Therefore, it is conceivable that in these cells mechanisms exist to metabolise this chemical agent. This would lead to a transient instead of a constant stimulus affecting the cells.

To test this hypothesis, primary hepatocytes have been stimulated with recombinant Wnt3a (rWnt3a) instead of SB216763. rWnt3a is the inartificial pathway activator that is not affected by mechanisms of detoxification. The temporal behaviour of β-catenin has been measured upon administration of this stimulus (data not shown; (Götschel, 2008)). Various dynamical characteristics differ from the dynamical properties detected after application of SB216763.

Upon Wnt stimulation, β-catenin increases much faster than after stimulation with SB216763. The β-catenin maximum is reached at ~2 h. The concentration increase is about 14-fold. A decreasing phase occurs. After 22 h, the concentration of β-catenin does not return to its unstimulated steady state value which is also in contrast to the β-catenin dynamics detected upon administration of the GSK3 inhibitor. The concentration stays higher. It is still about 8-fold increased compared to the unstimulated steady state level. Hence, the decreasing phase shows other characteristics if the pathway is activated by the inartificial instead of the artificial ligand.

We use the mathematical model and ask whether it is possible to distinguish between a decreasing phase that is caused by the degradation of the stimulus and a decreasing phase induced by the feedback loop. Such a differentiation will provide a deeper understanding of the underlying mechanisms of the decreasing phase in primary hepatocytes. For the further theoretical investigation of this question, we consider the two plausible extreme scenarios: i) the β-catenin concentration decreases only because of the Axin-2 feedback loop, and ii) the decrease is determined exclusively by the degradation of the stimulus. For scenario ii), the Axin-2 feedback loop is interrupted. The transient stimulus is modelled as an exponential decay in the Wnt concentration. In general, the hepatocyte-like model is used for the simulations as in this cell-type the deviation between model and experiment emerges.

In both scenarios, the β-catenin dynamics is simulated if a stimulus is applied at two successive time points (ts_1 and ts_2, $ts_2 > ts_1$). The second stimulus is given during the decreasing phase following the first stimulus. The temporal behaviour of the stimuli as well as the β-catenin dynamics are shown in Figure 2.38. In Figure 2.38A and Figure 2.38B, the dynamics simulating scenario i) and scenario ii), respectively, are shown.

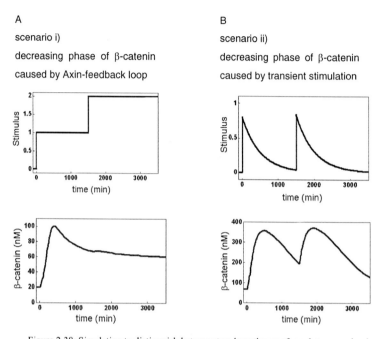

Figure 2.38: Simulation to distinguish between two hypotheses of regulatory mechanisms.
(A) The dynamics of the stimulus (upper part) and β-catenin (lower part) assuming scenario i) are presented. The feedback loop is interrupted. To this end, the production of the Axin-2 mRNA (k_{18}) is set to zero. (B) The dynamics assuming scenario ii) are shown. The transient stimulus is modelled as exponential decay of the Wnt concentration $Wnt(t) = 0.8 \cdot e^{-\lambda(t-ts_1)} + 0.8 \cdot e^{-\lambda(t-ts_2)}$, with $\lambda = 0.002$ min^{-1}, $ts_1 = 0$ min and $ts_2 = 1500$ min.
For the simulations of both settings, the hepatocyte-like model is used. The stimuli are applied at $t = 0$ min and $t = 1500$ min.

Figure 2.38 reveals that the β-catenin dynamics in the two scenarios differ. In scenario i), in which the feedback causes the decrease, the constant second stimulus is just added to the first one (Figure 2.38A). Due to the first stimulus, the β-catenin concentration increases as seen in Figure 2.37A. The feedback loop is switched on, leading to the decrease of the β-catenin concentration. During this decreasing phase, the second stimulus is applied. Until the second stimulus is given, the β-catenin concentration decreases from ~100 nM at the first maximum to ~66.9 nM. The second stimulus is not able to induce a strong increase of the β-catenin concentration. While the concentration at the first maximum is approximately 5-fold higher than the unstimulated steady state concentration, the β-catenin concentration at the second maximum is 67.6 nM, and thus, just slightly increased compared to the concentration at ts_2. This is due to the fact that the feedback loop is strongly activated because of its induction by the increased β-catenin concentration resulting from the first pathway activation. Hence, in this scenario the second stimulus causes a much weaker effect on the increase of the β-catenin concentration than the first stimulus. In addition to the magnitude of the two β-catenin maxima, their timing in relation to the respective pathway stimulation is considered. The second stimulus is reached approximately 100 min after application of the second stimulus. In contrast, the first maximum is reached ~400 min after the first pathway activation.

In Figure 2.38B, the dynamics of the stimulus and β-catenin are shown if scenario ii) is taken into account. As no negative feedback exists in this scenario, the β-catenin steady state is higher than in scenario i) in which the feedback via Axin-2 is included. In scenario ii), the unstimulated steady state concentration is ~68.7 nM. The first stimulus induces an increase of the β-catenin concentration. The β-catenin maximum is ~355 nM and reached at $t \sim 486$ min. Compared to scenario i), the maximum is reached ~86 min later. In scenario ii), the stimulus is degraded leading to the decrease of the concentration of β-catenin. Between the β-catenin maximum and the application of the second stimulus at ts_2, the β-catenin concentration decreases from ~355 nM to ~194 nM. The second stimulus causes a strong increase of the β-catenin concentration. Since no feedback loop exists in the model, the system reacts to the second stimulus as to the first stimulation. The β-catenin concentration strongly increases, reaching a maximum that is even slightly higher than the maximum reached upon the first pathway activation. The concentration rises to ~369 nM. The second β-catenin maximum is reached at $t \sim 1900$ min, 400 min after application of the second stimulus.

In summary, the simulations show that in scenario i), the second stimulus induces only a slight increase of the β-catenin concentration, while in scenario ii), the second stimulus causes a strong increase. In scenario ii) the second maximum is reached later than in scenario i). In consequence, the simulations result in the prediction that one is able to distinguish between the two hypotheses with an experiment in which the cells are stimulated at two successive time points. The second pathway stimulation has to be applied during the decreasing phase following the first stimulation.

The experiments suggested by the modelling results have been performed using primary hepatocytes. The dynamics are shown in Figure 2.39.

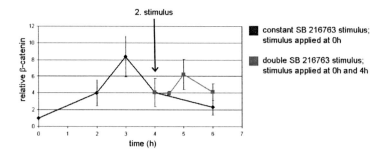

Figure 2.39: β-Catenin time course in a double stimulation experiment.
Primary hepatocytes have been treated with SB216763. The first stimulus has been administered at $t = 0$ h. At 4 h, the cell population has been split. One part has not been stimulated again staying with the single stimulation (constant stimulus, blue line). The other cell population has been additionally stimulated at $t = 4$ h (green line). The β-catenin concentration has been measured and the relative concentration change is plotted. The concentrations are related to the concentration in the unstimulated case at $t = 0$ h. The mean of three experiments and their standard deviation are shown.

The experiments show that the first stimulus induces an increase of the β-catenin concentration (Figure 2.39). After ~3 h, a maximum is reached which is about 8-fold higher than the unstimulated steady state concentration at $t = 0$ h. Thereafter, the β-catenin concentration decreases. During the decreasing phase, the second stimulus is applied at $t = 4$ h. Between the first maximum and the application of the second stimulus, the β-catenin concentration decreases to a relative level ~4. The second stimulus causes a second increase of the β-catenin concentration which does not start immediately after the treatment. Between 4 h and 4.5 h, almost no concentration change has been measured. The second maximum emerges around 5 h. The second maximum is not as high as the first maximum, but a clear increase of the relative β-catenin concentration is detectable. The relative concentration of the second maximum is ~6. The second peak is followed by a second decreasing phase.

The second stimulus is able to induce a strong increase of the β-catenin concentration. This indicates that the first decreasing phase is not induced exclusively by the Axin-2 feedback. As the second peak is not as high as the first one, this is an indication that the decrease is also not completely caused by a degradation of the added stimulus. Consequently, a combination of both scenarios is reasonable.

Please, note that the β-catenin dynamics shown in Figure 2.27 are slower than the β-catenin dynamics induced by the single constant stimulation shown in Figure 2.39 (blue line). In Figure 2.39, the first maximum is reached about 3 h after application of the first stimulus. The β-catenin dynamics presented in Figure 2.27 show the maximum at ~6 h. All experiments have been followed the same operating procedure. The experiments with primary hepatocytes show a high variation which is also indicated by the large error bars of the experimental data. This is a common issue occurring in experiments with primary cells. This might be even increased by the fact that hepatocytes are no homogeneous cell population. Depending on their location within the tissue they show differences (Braeuning et al., 2006; Gebhardt et al., 2007).

2.4.6   Sensitivity analysis

In section 2.4.4, the effects of the feedback strength, the total APC and the total E-cadherin concentrations on specific signalling characteristics are analysed. By these analyses, the combined effects of the model two extensions as well as the APC level are investigated. The impact of other model parameters is not investigated in these analyses. Section 2.4.6 concentrates on the impact of all parameters of the model on the unstimulated β-catenin steady state and the increasing phase. To this end, a sensitivity analysis is performed. The calculated sensitivity coefficients give insights about the type and the strength of impact of a reaction or concentration on the regarded characteristics (for details of calculation see section 2.3.4). For the calculation, the parameters are changed by 1%. The higher the value of the coefficient, the higher the impact of the particular reaction or concentration is. A positive value denotes that an increase of the parameter causes a higher β-catenin steady state concentration or signalling characteristics. The calculated sensitivity coefficients are presented as bar plots.

First, the sensitivity coefficients of the unstimulated β-catenin steady state of the C17.2-cell-like model and the hepatocyte-like model are compared to the sensitivity coefficients of the extract model. How do the model modifications affect the impact of the reactions on the steady state? Furthermore, the sensitivity coefficients of measures characterising the increasing phase are calculated. This analysis is performed only for the hepatocyte-like model. Experiments show that in these cells the increasing phase is very long. Is this long-lasting concentration increase only caused by the high concentration of E-cadherin? Which other parameters or concentrations have a strong effect on the increasing phase in this cell-type? Three measures to describe properties of the β-catenin dynamics are introduced and their sensitivity coefficients are compared. Last, the time-varying sensitivity coefficients are calculated in the extract and in the hepatocyte-like model. With this approach it is investigated how the impact of the parameters on the β-catenin concentration changes over time. Are there processes whose impact is strong (weak) immediately after pathway stimulation but weak (strong) at later time points? Do processes change their type of effect over time?

Sensitivity coefficients of the unstimulated steady state concentration of β-catenin

The effects of parameter changes on the unstimulated steady state of β-catenin are analysed. The sensitivity coefficients of this property are shown in Figure 2.40.

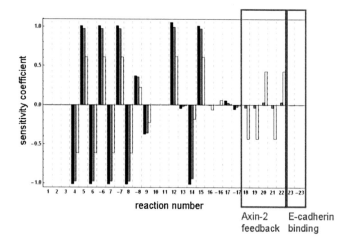

Figure 2.40: Comparing sensitivity coefficients of the unstimulated steady state of β-catenin in the extract, the C17.2-cell-like and the hepatocyte-like model.
The sensitivity coefficients of the extract model (black bars), the C17.2-cell-like model (grey bars) and the hepatocyte-like model (white bars) are shown. The parameters are changed by 1%. The numbers correspond to the reactions numbers shown in the model scheme in Figure 2.29. In the case of reversible reactions, the complex formation is denoted with a positive sign; the dissociation of the complex is labelled with the negative reaction number.
Reactions of the Axin-2 feedback loop are framed in red, the reactions including E-cadherin are framed in green.

In all three models, the impact of the core-reactions (reactions 1 to –17) stays high (Figure 2.40). The reactions of the destruction cycle and the formation of the destruction complex (reactions 4 to 9) are of high importance even in the presence of the feedback loop. However, their impact decreases with the inclusion of the feedback loop via Axin-2. In the hepatocyte-like model in which the feedback loop is strong (white bars), their impact is lower than in the C17.2-cell-like model, in which the feedback loop is weak (grey bars). In general it holds that the weaker the feedback loop is, the stronger the impact of the core-reactions on the unstimulated β-catenin steady state. The impact of the constant, feedback-loop-independent production of Axin (reaction 14) shows the strongest change of the strength of effect. In the cellular models, Axin is produced in two ways: i) by the constant production, and ii) by the production that is induced by the intermediate (reaction 19). The impact of the constant production of Axin decreases since the overall impact of the Axin production is partially taken over by the inducible Axin-2 production.

There is one exception of the general description that the impact of the core-reactions decreases with increasing feedback strength. The impact of the formation and dissociation of the TCF/β-catenin complex (reactions 16 and –16, respectively) increases. This is due to the fact that β-catenin bound to TCF cannot be degraded and it induces the feedback loop. Therefore, the formation of the TCF/β-catenin complex has an effect on the concentration of β-catenin. The impacts of all other reactions of the feedback loop (reactions 18 to 22) also increase with increasing strength of the feedback loop (compare grey and white bars, framed in red in Figure 2.40). Reactions facilitating the production of Axin have negative effects (reactions 18, 19 and 21) and reactions that reduce the Axin production have positive sensitivity coefficients (reactions 20 and 21).

As already shown in the simulations above (Figure 2.30, section 2.4.3), the formation and dissociation of the E-cadherin/β-catenin complex, do not affect the unstimulated steady state concentration of β-catenin (reactions 23 and –23, framed in green in Figure 2.40).

Characterisation of the timing of the β-catenin maximum

In section 2.4.4, the combined effect of the feedback strength and varying total APC and E-cadherin concentrations on t_{max} is examined. Here, the impact of all kinetic parameters of the model on the increasing phase is investigated. Moreover, the impact of the other protein concentrations whose total levels do not change over time, namely Dsh, TCF and GSK3, is examined. Furthermore, the impact on t_{max} will be compared to the impact on two other measures describing the β-catenin dynamics. To this end, the sensitivity coefficients of all three measures are calculated.

First, a measure characterising the increasing phase is introduced which is referred to as increase duration τ_{inc}. This is the area below the β-catenin curve between the time the stimulus is given (t_{stim}) and t_{max}. It is illustrated in Figure 2.41 (orange area) and defined by

$$\tau_{inc} = \frac{\int_{t_{stim}}^{t_{max}} t \cdot (\beta\text{-}catenin[t] - \beta\text{-}catenin^{unstim}) dt}{\int_{t_{stim}}^{t_{max}} (\beta\text{-}catenin[t] - \beta\text{-}catenin^{unstim}) dt} \tag{2.13}$$

where β-$catenin^{unstim}$ is the unstimulated steady state concentration of β-catenin.

The measures t_{max} and τ_{inc} characterise the increasing phase. Additionally, the characteristic signalling time t_{Llo} introduced by Llorens and colleagues (Llorens et al., 1999) is of interest. In contrast to t_{max} and τ_{inc}, t_{Llo} considers the increasing and the decreasing phase. Therefore, t_{Llo} is a measure to describe the overall β-catenin dynamics. It is included in the analysis to compare the effects on the increasing phase to the effects on the overall dynamics. The area representing t_{Llo} is shaded in grey in Figure 2.41. It is calculated by:

$$t_{Llo} = \frac{\int_{0}^{\infty} t \cdot |(\beta\text{-}catenin'[t])| dt}{\int_{0}^{\infty} |(\beta\text{-}catenin'[t])| dt} \tag{2.14}$$

Figure 2.41 shows a graphical description of the three measures.

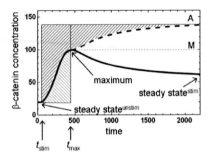

Figure 2.41: Measures to characterise the timing of the β-catenin maximum.

Graphical representation of t_{max}, the increasing duration (τ_{inc}) (orange area) and the characteristic time introduced by Llorens and colleagues (t_{Llo}) (grey area) (Llorens et al., 1999).

The grey area is generated the following: from t_{stim} to t_{max}, the solid blue line is considered and for $t > t_{max}$ the dashed blue line is taken into account. This is the mirror image of the solid line with respect to the dotted grey line M. The area between the asymptotic state A and the solid line between t_{stim} and t_{max} as well as the dashed line from t_{max} to ∞ is considered.

The impact of the reactions on these three time characteristics is analysed in the hepatocyte-like model, shown in Figure 2.42. The analysis demonstrates that the types of effect (signs of the sensitivity coefficients) of the individual reactions are equal for all three characteristics t_{max}, τ_{inc} and t_{Llo}. The only exception is the Dsh-induced release of GSK3 from the complex APC/Axin/GSK3 (reaction 3). However, its impact on all three characteristics is very small.

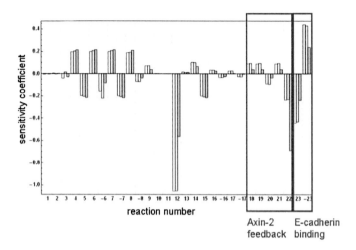

Figure 2.42: Sensitivity coefficients of the time characteristics.
The sensitivity coefficients of t_{max} (white bars), τ_{inc} (orange bars) and t_{Llo} (grey bars) in the hepatocyte-like model are shown. The parameters are changed by 1%. The numbers correspond to the reactions numbers shown in the model scheme in Figure 2.29. In the case of reversible reactions, the complex formation is denoted with a positive sign; the dissociation of the complex is labelled with the negative reaction number. The reactions of the feedback loop are framed in red while the reactions containing E-cadherin are green-framed.

Regarding τ_{inc} and t_{max}, the strength of impact of all reactions is very similar (white and orange bars, respectively, in Figure 2.42). In contrast, for various reactions the coefficients with respect to t_{Llo} differ clearly in strength (grey bars in Figure 2.42). This is caused by the fact that t_{Llo} includes the decreasing phase.

With respect to τ_{inc} and t_{max}, the production rate of β-catenin (reaction 12) has the strongest effect (reaction 12). The reactions of the formation and dissociation of the E-cadherin/β-catenin complex (reactions 23 and −23, framed in green) have the second strongest impact. The sensitivity coefficient of the association (reaction 23) is negative. Thus, an increase of this rate decreases the signalling characteristic. The increasing phase is reduced. At the first sight, this is counterintuitive as one would expect that a stronger binding to E-cadherin leads to a longer lasting buffering effect. But the opposite is the case in a system with constant total E-cadherin concentration. If the rate constant of the binding reaction of β-catenin and E-cadherin is larger, the pool of unbound E-cadherin is lower in the unstimulated case. Consequently, less E-cadherin is available to bind β-catenin if a stimulus is applied. The buffering effect is lower and the β-catenin concentration increases faster (shortened increasing phase) compared to the model in which the binding of E-cadherin and β-catenin is weaker.

The reactions of the feedback loop (reactions 18 to 22; framed in red) as well as the core-reactions (reactions 1 to −17) participate in the regulation of the timing of the increasing phase. Exceptions are the release of phosphorylated β-catenin from the destruction complex (reaction 10) and the degradation of the phosphorylated β-catenin (reaction 11). They do not affect the timing.

Regarding t_{Llo}, the impact of the production rate of β-catenin (reaction 12) is high, but not as high as its impact on t_{max} and τ_{inc}. The degradation of the intermediate (reaction 22) has the strongest impact on t_{Llo}. An increase of this degradation rate enhances the increasing phase but also the decreasing phase, causing a faster reaching of the stimulated β-catenin steady state. Therefore, the impact on t_{Llo} is very strong. Like for τ_{inc} and t_{max}, the reactions involving E-cadherin (reactions 23 and −23) influence t_{Llo} strongly. As binding to E-cadherin affects the increasing as well as the decreasing phase, the sensitivity coefficients concerning t_{Llo} are smaller than the sensitivity coefficients with respect to t_{max} and τ_{inc}. Furthermore, the reactions of the feedback loop (reactions 18 to 22) and the core-reactions influence t_{Llo}.

In addition to the analysis of the effects of the individual reactions, the sensitivity coefficients of τ_{inc}, t_{max} and t_{Llo} are calculated with respect to the concentrations of TCF, GSK3 (GSK in Figure 2.43), APC, Dsh and E-cadherin (Figure 2.43). These are the proteins obeying conservation relations in the model.

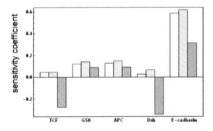

Figure 2.43: Sensitivity coefficients of the time characteristics of the total protein concentrations.
The regarded signalling characteristics are: t_{max} (white bars), τ_{inc} (orange bars) and t_{Llo} (grey bars). The parameter set of the hepatocyte-like system is used. The concentrations are changed by 1%.

The results in Figure 2.43 show that all total concentrations affect all three considered signalling characteristics. The concentration changes of TCF and Dsh affect t_{max} and τ_{inc} in the same but t_{Llo} in the opposite way. Changes in the levels of GSK3, APC and E-cadherin have the same type of effect on all three time characteristics. However, the strength of impact is different.

GSK3, APC and E-cadherin have a positive influence on all three time measures. The phase of an increasing β-catenin concentration is prolonged. The sensitivity coefficients of TCF and Dsh with respect to t_{max} and τ_{inc} are positive. Concerning t_{Llo} they are negative. This is reasoned by the fact that t_{Llo} takes into account the increasing and the decreasing phase. A higher concentration of TCF prolongs the increasing phase. TCF/β-catenin is part of the feedback loop. The consequence of a higher concentration of TCF is a stronger feedback loop, and thus, a lower β-catenin concentration. Therefore, the buffering effect of E-cadherin is stronger and β-catenin does not increase as fast as under conditions of a lower TCF concentration. The maximum is reached later than in the unperturbed model. The decreasing phase of β-catenin is flattened compared to the system's behaviour simulated with the original concentrations (dynamics not shown). Hence, the part of the decreasing phase on t_{Llo} is smaller. Thus, although the increasing phase lasts longer than in the system with original concentrations, t_{Llo} is smaller as the decreasing phase is shortened. In the case of Dsh, the differences in t_{max} and τ_{inc} to t_{Llo} can be explained the following: The more Dsh exists in the model, the stronger the stimulus is. Consequently, β-catenin increases to a higher level than in the case with a lower Dsh concentration. Although the increase is quite fast, it takes longer to reach the β-catenin maximum; t_{max} and τ_{inc} increase. As β-catenin increases to a higher level, the feedback loop is stronger activated. Therefore, the decreasing phase of β-catenin is reduced, causing the negative impact of Dsh on t_{Llo}.

E-cadherin has the strongest positive effect on t_{max}, t_{Llo} and τ_{inc}. The more E-cadherin is available in the system, the stronger the stimulus-induced increase of the β-catenin concentration is buffered (see also Figure 2.35). For all three measures, the sensitivity coefficients of APC and GSK3 are positive. The more GSK3 or APC exists, the more destruction complex is built and the lower the concentration of β-catenin is. This enables a stronger buffering effect of E-cadherin. The characteristics of the timing increase.

In summary, the analyses show that the sensitivity coefficients of t_{max} and τ_{inc} are very similar. This holds for the sensitivity coefficients of the reactions and the sensitivity coefficients of the total protein concentrations. Thus, to that effects, t_{max} describes the increasing phase well. It is not necessary to always calculate τ_{inc}.

The effects on the characteristic time t_{Llo} show clear differences in comparison to the effects on t_{max} and τ_{inc} since t_{Llo} takes the increasing and the decreasing phase into account. Concerning the sensitivity coefficients of the reactions (Figure 2.42), the reactions have the same type of effect for all three measures. However for some reactions, the strength of impact strongly differs between a measure exclusively characterising the increasing phase and the measure taking the overall dynamics into account. These reactions whose impact on t_{Llo} is weaker than on t_{max} and τ_{inc} are the Dsh-induced release of GSK3 from the complex APC/Axin/GSK3 (reaction −6), the β-catenin synthesis (reaction 12), the formation and dissociation of the E-cadherin/β-catenin complex (reactions 23 and −23) and the reactions of the feedback loop (reactions 18 to 22) with the exception of the degradation of the intermediate (reaction 22). Its impact is strongly increased if t_{Llo} is considered. These differences are based on the inclusion of the decreasing phase in the calculation of t_{Llo}. Hence, the impact concerning t_{Llo} allows concluding how the reactions affect the decreasing phase. This also holds for the impact of the protein concentrations whose total concentrations do not change over time (Figure 2.43).

Time-dependent sensitivity analysis

The sensitivity coefficients calculated above quantify the effect of one reaction on the β-catenin steady state or on measures characterising the temporal behaviour of β-catenin. Additionally, we are interested in how the single reactions affect the β-catenin concentration over time. Do all reactions show their maximal influence immediately after pathway stimulation? Is the main impact on the β-catenin concentration mediated by different reactions at different time point? Are there any reactions whose type of impact (positive/negative) changes over time?

To address these questions the time-varying sensitivity coefficients are calculated (Ingalls and Sauro, 2003). This method allows tracing the impact of a parameter perturbation on the readout over time. The non-normalised time-varying sensitivity coefficients of concentration $R_q^s(t)$ are calculated by

$$R_q^s(t) := \frac{\partial s(t,q)}{\partial q}\bigg|_{q=q_0} \quad \text{for } t \geq 0. \tag{2.15}$$

Here, s is the vector of the concentrations of the pathway components and q a vector containing the initial conditions and external parameters. The standard set of initial conditions and initial external parameters compose the initial vector q_0.

The time-varying concentration sensitivity coefficients may change the sign over time. This means that their type of impact changes over time. Also the strength of impact can change over time. If the stimulated steady state is reached, the time-varying concentration sensitivity coefficients are equal to the sensitivity coefficients calculated for the stimulated steady state.

We investigate the time-varying sensitivity coefficients of β-catenin in the extract model (Figure 2.44) and the hepatocyte-like model (Figure 2.45). The coefficients are presented if their absolute maximal value is larger than 0.1. The coefficients are grouped with respect to magnitude and sign: (A) large positive values, (B) large negative values, (C) intermediate positive and (D) intermediate negative values. In (E) the coefficients changing their sign are presented. In the extract model, the absolute values of the coefficients of the following reactions are below the threshold of 0.1: i) the activation of Dsh_i (reaction 1), ii) the deactivation of Dsh_a (reaction 2), iii) the release of phosphorylated β-catenin from the destruction complex (reaction 10), iv) the degradation of phosphorylated β-catenin (reaction 11), and v) the formation and dissociation of the TCF/β-catenin complex (reactions 16 and –16, respectively).

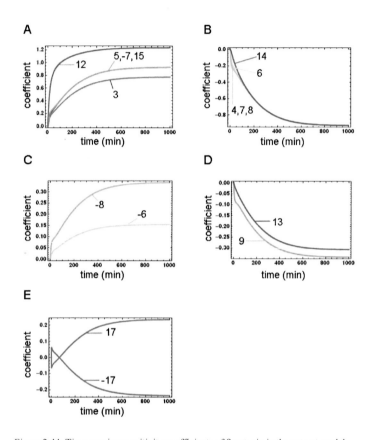

Figure 2.44: Time-varying sensitivity coefficients of β-catenin in the extract model.

For the extract model, the time-varying sensitivity coefficients of unbound, unmodified β-catenin are calculated. The reaction numbers of the particular reaction (according to the model scheme shown in Figure 2.29) are written at the respective curve. In the case of reversible reactions, the complex formation is denoted with a positive sign; the dissociation of the complex is labelled with the negative reaction number. (A) The coefficients with large positive values are shown. These are: i) the Dsh-dependent dissociation of the APC/Axin/GSK3 complex (reaction 3), ii) the dephosphorylation of APC and Axin bound to GSK3 (reaction 5), iii) the dissociation of the APC/Axin complex (reaction −7), iv) the synthesis of β-catenin (reaction 12), and v) the degradation of Axin (reaction 15). (B) The coefficients with large negative values are presented. They relate to i) the phosphorylation of APC and Axin bound to GSK3 (reaction 4), ii) the binding of GSK3 to APC/Axin (reaction 6), iii) the binding of APC and Axin (reaction 7), iv) the binding of β-catenin to the destruction complex (reaction 8) as well as v) the production of Axin (reaction 14). In (C) the coefficients with positive and in (D) the coefficients with (continued on page 117)

The majority of processes has a maximum of the time-varying sensitivity coefficients whose absolute value is larger than 0.1 (Figure 2.44). The production of β-catenin (reaction 12), the degradation of Axin (reaction 15) as well as the reactions leading to the decomposition of the destruction complex (reactions 3, 5, –6, –7 and –8) have positive time-varying sensitivity coefficients (Figure 2.44A and Figure 2.44C). The impact of these reactions increase over time until β-catenin reaches its stimulated steady state level. Then the time-varying sensitivity coefficients stay constant. They are equal to the sensitivity coefficients of the stimulated system (the types of effect are comparable with the results presented in Figure 2.5, section 2.2). The reactions facilitating the production of the destruction complex (reactions 4, 6, 7 and 8), the passing through the destruction cycle (reactions 9 and 10), the production of Axin (reaction 14) as well as the alternative β-catenin degradation have negative time-varying sensitivity coefficients (Figure 2.44B and Figure 2.44D). This type of effect stays the same over time.

There are two reactions whose sensitivity coefficients change their signs over time: i) the binding of APC and β-catenin (reaction 17), and ii) the dissociation of the APC/β-catenin complex (reaction –17). Immediately after pathway activation, the formation of the APC/β-catenin has a negative impact on the concentration of unbound β-catenin. APC binds β-catenin, lowering the concentration of unbound β-catenin. In the following, the effect changes its sign and the binding reaction has a positive impact as it decreases the pool of APC that is available for the formation of the destruction complex. Less destruction complex is built and consequently, less β-catenin is degraded via the destruction cycle (as also discussed in sections 2.2, 2.3, and 2.4.4). For the dissociation of the APC/β-catenin complex (reaction –17), the line of argument is the opposite way.

Continued legend of Figure 2.44.

intermediate negative values are shown. In (C), these are the coefficients of the Dsh-independent dissociation of APC/Axin/GSK3 (reaction –6) and the release of β-catenin from the destruction complex (reaction –8). In (D), the coefficients of the phosphorylation of β-catenin (reaction 9) and the alternative β-catenin degradation (reaction 13) are shown. (E) The coefficients changing the type of effect over time are presented. These are the formation and dissociation of the APC/β-catenin complex (reactions 17 and –17, respectively).

With respect to predictions of experiments, reactions whose sensitivity coefficients change their sign over time could be of particular interest. For this model one can predict that activating the dissociation of the APC/β-catenin complex leads to a higher β-catenin level immediately after pathway stimulation, but from $t > \sim 65$ min to a lower β-catenin concentration compared to the system in which the dissociation is not activated. As discussed in sections 2.2 and 2.3 the experimental verification of this theoretical prediction is complicated as the exclusive perturbation of the formation or dissociation of the APC/β-catenin complex is difficult.

In Figure 2.45, the time-varying sensitivity coefficients calculated in the hepatocyte-like system are presented. The absolute sensitivity coefficients of the following reactions do not reach a value larger than 0.1: the activation of Dsh_i (reaction 1), the deactivation of Dsh_a (reaction 2), the release of phosphorylated β-catenin from the destruction complex (reaction 10), the degradation of phosphorylated β-catenin (reaction 11), the alternative β-catenin degradation (reaction 13), the formation and dissociation of the TCF/β-catenin complex (reactions 16 and −16, respectively) and the formation and dissociation of the APC/β-catenin complex (reactions 17 and −17, respectively). Thus, they are not presented in Figure 2.45.

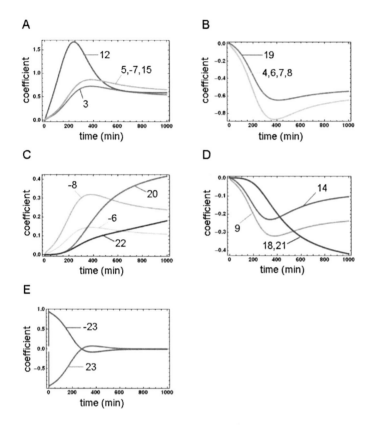

Figure 2.45: Time-varying sensitivity coefficients of β-catenin concentration in the hepatocyte-like system.

The time-varying sensitivity coefficients of β-catenin are calculated. The reaction numbers of the particular reaction (according to the model scheme shown in Figure 2.29) are written at the respective curve. In the case of reversible reactions, the complex formation is denoted with a positive sign; the dissociation of the complex is labelled with the negative reaction number. (A) The coefficients with large positive values are shown. They are related to the following reactions: i) the Dsh-dependent dissociation of the APC/Axin/GSK3 complex (reaction 3), ii) the dephosphorylation of APC and Axin bound to GSK3 (reaction 5), iii) the dissociation of the APC/Axin complex (reaction −7), iv) the synthesis of β-catenin (reaction 12) and v) the degradation of Axin (reaction 15). (B) The coefficients with large negative values are presented. They relate to i) the phosphorylation of APC and Axin bound to GSK3 (reaction 4), ii) the binding of GSK3 to APC/Axin (reaction 6), iii) the binding of APC and Axin (reaction 7), iv) the binding of β-catenin to the destruction complex (reaction 8), and v) the intermediate-dependent production of (continued on page 120)

The plots in Figure 2.45 display that the impact of the particular parameters on the β-catenin concentration changes over time. At specific time points different parameters contribute to a distinct extent to the "overall impact" on β-catenin.

Immediately after pathway activation, the formation and dissociation of the E-cadherin/β-catenin complex (reactions 23 and −23, respectively) mainly determine the β-catenin concentration (Figure 2.45E). This emphasises the influence of E-cadherin during the increasing phase as a buffer of β-catenin (discussed before in sections 2.4.2, 2.4.4, and 2.4.6). In the beginning, binding to E-cadherin (process 23) has a negative effect on the β-catenin level as the pool of unbound β-catenin is lowered. Afterwards, this impact decreases and around $t = 300$ min, the impact changes its sign. The type of effect changes. At later time points and if β-catenin reaches its stimulated steady state, the formation and dissociation of the E-cadherin/β-catenin complex has no impact of the unbound β-catenin (see also section 2.4.3, Figure 2.30). The small transient positive effect of the binding of E-cadherin and β-catenin at intermediate time points (between ~280 min and ~700 min) results from the feedback loop. If the feedback loop is active, an increase in the β-catenin concentration also causes a stronger activation of the feedback loop. In consequence, an increase of a parameter resulting in an increase in the β-catenin level has two effects: i) It increases the β-catenin concentration, but ii) at the same time it stronger activates the feedback loop, leading to a

---

Continued legend of Figure 2.45.

Axin (reaction 19). In (C) the coefficients with intermediate positive and in (D) the coefficients with intermediate negative values are shown. In (C) these are the coefficients related to the Dsh-independent release of GSK3 from the complex APC/Axin/GSK3 (reaction −6), the release of β-catenin from the destruction complex (reaction −8) as well as the degradation of the Axin-2 mRNA (reaction 20) and of the intermediate (reaction 22). In (D) the coefficients of the phosphorylation of β-catenin (reaction 9), the constant Axin production (reaction 14) and the production of the Axin-2 mRNA (reaction 18) and the intermediate (reaction 22) are shown. In (E) the coefficients changing their type of effect over time are presented. These are the formation and the dissociation of the E-cadherin/β-catenin complex (reactions 23 and −23, respectively).

decrease of the β-catenin concentration. This is also the reason why the absolute effects of several parameters do not only increase until they reach their values at the steady state but also decrease at later time points. The effect is developed the strongest in the case of the β-catenin production (reaction 12; Figure 2.45A). Upon pathway stimulation, its impact increases and even reaches a higher value than in the extract model (Figure 2.44A). The higher the β-catenin production is, the more β-catenin is produced, leading to the strong positive impact. Later, an increase of this parameter does not only positively affect the β-catenin concentration since the feedback loop is stronger activated and more β-catenin can be degraded via the destruction cycle. Hence, the impact of the β-catenin production decreases and achieves a lower value than in the extract model. In general, the same holds for the other parameters apart from the production and degradation of the Axin-2 mRNA (reactions 18 and 20, respectively) and of the production and degradation of the intermediate (reactions 21 and 22, respectively). These reactions are part of the feedback loop. Regarding the absolute value of the sensitivity coefficients, their values increase over time till they reach the maximal value if the stimulated steady state of β-catenin is reached (not shown). The feedback loop increases its control on the β-catenin concentration over time.

The comparison of the time-varying sensitivity coefficients in the hepatocyte-like model (Figure 2.45) and the extract model (Figure 2.44) displays that the feedback loop decreases the control of the core-reactions (reactions 1 to -17) on the β-catenin concentration over time. This is also shown for the unstimulated steady state (Figure 2.40). In the hepatocyte-like as well as in the extract model, a switch of the type of effect is observable. In both systems, the impact of the formation and dissociation of the APC/β-catenin complex (reactions 17 and -17) change their signs over time. In the extract model, the absolute values of the coefficients are larger than 0.1. However, in the hepatocyte-like system their impacts are smaller and therefore not presented in Figure 2.45. In addition, in the hepatocyte-like model, the coefficients of the formation and dissociation of the E-cadherin/β-catenin complex change their signs over time.

2.4.7   Effects of APC mutations

In the next theoretical study, we are interested in the effects of APC mutations in the cellular models. In their publication on MAPK signalling, C. Kiel and L. Serrano have shown that mutations have different effects on the temporal behaviour, depending on the strength of the feedback loop (Kiel and Serrano, 2009): Mutations within the feedback loop have stronger effects on the temporal behaviour if the feedback loop is weak but very little effects in cases of a strong feedback loop.

In this section we address the question how distinct APC mutations (introduced in section 2.2, (Cho et al., 2006)) affect the β-catenin dynamics in the hepatocyte-like and in the C17.2-cell-like model. We investigate the effects of the APC mutations m1, m5 and m10 in these models. The effects of the APC mutations in the extract model are investigated in section 2.2.3. Here, we ask whether APC mutations cause similar or distinct effects depending on the kind of cellular system. How do APC mutations of different strength influence the β-catenin behaviour? Moreover, we ask whether inhibitors work efficiently in both cellular models or whether there are differences depending on the kind of cellular model and the strength of the occurring APC mutation.

Effects of different APC mutations on the β-catenin dynamics in the hepatocyte-like model and the C17.2-cell-like model

For the two cellular models, we analyse the temporal behaviour of β-catenin if APC is mutated. We compare the β-catenin dynamics in the hepatocyte-like system and the C17.2-cell-like system under wild-type conditions and conditions of the APC mutations m1, m5 and m10. The parameters of the APC mutations are adapted to the publication by Cho and colleagues (Cho et al., 2006) and are provided in Table 6, Appendix A.

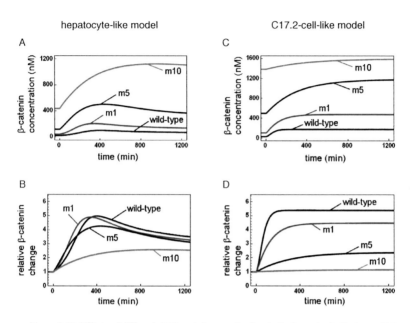

Figure 2.46: Effects of different APC mutations on the β-catenin dynamics in the two cellular models.

The β-catenin dynamics in (A and B) the hepatocyte-like and in (C and D) the C17.2-cell-like model are shown. (A) and (C) show the absolute β-catenin time courses. In (B) and (D), the relative β-catenin dynamics are shown. For the relative concentration changes the concentration is related to the respective unstimulated steady state concentration. At $t = 0$ min, the pathway is constantly activated by a Wnt stimulus.

The β-catenin dynamics under wild-type conditions (black line), APC mutation m1 (turquoise line), APC mutation m5 (blue line), and APC mutation m10 (cyan line) are shown. Parameters of the mutations are provided in Table 6, Appendix A.

Figure 2.46A shows the absolute concentration changes of β-catenin under wild-type conditions and conditions of weak (turquoise), intermediate (blue), or strong (cyan) APC mutations in the hepatocyte-like model. In Figure 2.46B their relative concentration changes are presented which are related to the respective unstimulated β-catenin steady state concentrations. Figure 2.46C and Figure 2.46D show the absolute and relative β-catenin dynamics, respectively, in the C17.2-cell-like model.

Due to the APC mutation, the unstimulated steady state levels of β-catenin increase in both models. Since the feedback is weaker in the C17.2-cell-like system, the β-catenin steady state concentrations in the C17.2-like model are higher than the corresponding β-catenin steady state concentrations in the hepatocyte-like model. In the extract system, the β-catenin steady states of the wild-type and the models with APC mutation are even higher than in the C17.2-cell-like system (Figure 2.8, section 2.2.3). Additionally, the β-catenin level is stronger increased, the stronger the mutation is (for the extract model see Figure 2.8, section 2.2.3). This is based on the fact that the stronger the mutation is, the more binding sites of APC are deleted, and the less β-catenin is degraded via the destruction cycle.

The pathway activation leads to an increase of the β-catenin concentration in both cellular models and in the absence and presence of an APC mutation. As the absolute values of the concentrations vary strongly, we also analyse the relative concentration changes in the two cellular models (Figure 2.46B and Figure 2.46D). The comparison of the relative concentration changes of β-catenin in the hepatocyte-like model (Figure 2.46B) in the wild-type case and mutant cases m1 and m5 does not reveal strong differences. In the wild-type and the m1 system, the ratio between the unstimulated steady state and the β-catenin maximum is ~5. In the m5 system, this ratio is about 4.25. In the wild-type model, the maximum is reached at ~402 min. In the m1 system it is achieved slightly earlier at ~395 min and in the m5 system slightly delayed at ~439 min. The β-catenin fold-change (ratio between the stimulated and unstimulated β-catenin steady state) is very similar in the wild-type, m1 and m5 models. In the wild-type model it is ~2.85, and slightly decreased in the m1 and m5 models where ratios of ~2.73 and ~2.58, respectively, are achieved. Hence, in the hepatocyte-like model, there are only slight differences in the relative β-catenin maximum, its timing and the fold-change between the wild-type, the m1 and the m5 model. The simulations of the m10 mutation in the hepatocyte-like model show some more pronounced differences compared to the three other models. The strong APC mutation m10 results in a lower relative β-catenin concentration change. The β-catenin concentration increases about 2.57-fold. Due to the strong mutation, the β-catenin concentration is strongly increased (see Figure 2.46A). The pathway activation is not able to induce a relative concentration change as strong as in the wild-type model or under conditions of weak or intermediate APC mutations. The β-catenin maximum is reached at ~950 min, and thus, ~550 min later than in the wild-type model. These are strong differences between the m10 and the wild-type model. In the m10 model, the fold-change is ~2.17 which is roughly one quarter less than the fold-change in the wild-type model.

Taken together the simulations of the hepatocyte-like model show that concerning the relative concentrations, the differences between the β-catenin dynamics in the wild-type, m1 and m5 models are very small. A strong APC mutation (m10) is needed to cause strong differences in the relative β-catenin dynamics.

In the C17.2-cell-like model, the relative temporal behaviour of β-catenin in the wild-type model differs from the behaviour under mutant conditions (Figure 2.46D). The stronger the APC mutation, the lower the relative β-catenin concentration changes are. Since in the C17.2-cell-like model no decreasing phase occurs, a separate examination of the fold-change and the relative β-catenin maximum is not necessary. The fold-change decreases with increasing strength of the APC mutation. While in the wild-type model the stimulus induces a ~5.4-fold concentration increase, the fold-change in the m1 model is ~4.5. In the m5 and m10 model, the concentration increases by a factor of ~2.4 and ~1.15, respectively. While in the wild-type case in both cellular models the maximal relative concentration changes are quite similar, they differ in the presence of an APC mutation. The maximal relative concentration changes under the m1 conditions in the C17.2-cell-like model and under the m5 conditions in the hepatocyte-like model are similar. The same holds for the maximal relative concentration changes if mutation m10 occurs in the C17.2-cell-like system and the m5 mutation in the hepatocyte-like system. As in the hepatocyte-like model, the increasing phase is analysed in the C17.2-cell-like model. For quantification, the time point is calculated at which 99.5% of the stimulated steady state concentration is reached. Under wild-type conditions, this is reached at ~227.4 min, while it takes ~1700 min to reach this characteristic concentration in the m1 model. In the m5 model, the increasing phase lasts ~3400 min and in the m10 model it lasts ~5050 min. Thus, in the C17.2-cell-like model it holds that the stronger the APC mutation is, the longer the increasing phase lasts. In the hepatocyte-like model, a prolongation of the increasing phase appears only if APC is strongly mutated.

In summary, the simulations of the C17.2-cell-like model show that the presence of an APC mutation decreases the β-catenin fold-change and prolongs the increasing phase. This property already appears for weak APC mutations. The stronger the mutation is, the stronger the difference is developed.

The comparison of both cellular models leads to the conclusion that with respect to the relative β-catenin dynamics, the hepatocyte-like model compensates for weak or intermediate APC mutations but not for strong APC mutations while the C17.2-cell-like model does not compensate for APC mutations of any strengths. Regarding the absolute β-catenin concentrations it holds for both cellular models that the stronger the APC mutation is, the higher the β-catenin levels.

Effects of inhibitions on the unstimulated β-catenin steady state in the cellular models with APC mutations

In part, different APC mutations cause distinct effects in the two cellular models (Figure 2.46). Subsequent we ask whether also inhibitions result in different effects depending on the cellular model and the occurring APC mutation. Does an inhibitor affect the unstimulated β-catenin steady state in the hepatocyte-like model similarly to this steady state in the extract model or in the C17.2-cell-like model? How are the effects of inhibitions in cellular models if an APC mutation occurs? The question is asked whether the same inhibitions are efficient as in the extract model or in a wild-type cellular models.

To analyse the effects of inhibitions, single rate constants are decreased by 50% and the effect on the unstimulated steady state of β-catenin is analysed (for details see section 2.2). The corresponding analysis focussing on the effects of inhibitions on the β-catenin fold-change is shown in Appendix C (Figure C.2). The effects in a cellular model without or with the APC mutations m1, m5 or m10 (Cho et al., 2006) as well as the effects in the extract model without and with an APC mutation are shown in Figure 2.47.

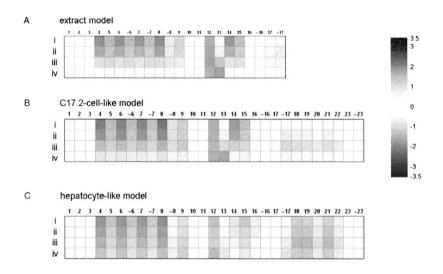

Figure 2.47: Effects of inhibitions on the unstimulated β-catenin steady state in the extract model and the cellular models in the presence and absence of APC mutations.

In (A) the analysis in the extract model is shown. In (B) and (C) the analyses of the C17.2-cell-like and the hepatocyte-like model, respectively, are presented. The effects of an inhibition of rate constants by 50% on the unstimulated β-catenin steady state are examined for the wild-type case (row i) and APC mutations m1 (row ii), m5 (row iii) and m10 (row iv). Parameters of the particular APC mutations are provided in Table 6, Appendix A.

White: no impact; green: an inhibition leads to a lower steady state; red: an inhibition leads to a higher steady state; the darker the colour, the higher the impact (see legend).

The numbers correspond to the reaction numbers shown in the model scheme in Figure 2.29. In the case of reversible reactions, the complex formation is denoted with a positive sign; the dissociation of the complex is labelled with the negative reaction number.

In Figure 2.47 it is demonstrated that for all three analysed models it holds that under wild-type conditions (row i) the destruction reactions (reactions 4 to 9, 14 and 15) as well as the β-catenin turnover (reactions 12 and 13) have a strong impact on the unstimulated β-catenin steady state. The impact of these reactions is very similar in the extract model (Figure 2.47A) and the C17.2-cell-like model (Figure 2.47B) and lower in the hepatocyte-like system (Figure 2.47C). The impact of the reactions of the feedback loop (reactions 18 to 22) is low in the C17.2-cell-like system. These reactions have a stronger effect in the hepatocyte-like system. For smaller perturbations these general observations are also seen in the sensitivity analysis shown in Figure 2.40, section 2.4.6.

The extract model is equal to the wild-type model in section 2.2. Hence, the effects of APC mutations in this model are already analysed in section 2.2.3 (Figure 2.9). Figure 2.47A and Figure 2.9 (row i to row iv) show the same information. For the sake of an easier comparison with the effects in the cellular models, it is shown again in Figure 2.47A. A short summary of the effects is given. The details are provided in section 2.2.3. In the extract model (Figure 2.47A), the impact of the destruction reactions (reactions 4 to 9, 14 and 15) decreases if APC is mutated. The stronger the APC mutation is, the lower the impact of the destruction reactions. In contrast, the impact of the alternative β-catenin degradation (reaction 13) increases strongly. This increase is caused by the fact that the stronger the APC mutation, the less β-catenin is degraded via the destruction cycle. In consequence, the influence of the destruction reactions decreases and the impact of the alternative β-catenin degradation increases.

This general observation also holds for the C17.2-cell-like model. The stronger the APC mutation is, the lower the impact of the destruction reactions (reactions 4 to 9, 14 and 15) and the higher the impact of the alternative β-catenin degradation (reaction 13). However, the decrease of the impact of the destruction reactions in the case of APC mutations is not as strong as in the extract model (compare rows ii to iv in Figure 2.47A and B). In the hepatocyte-like model, the decrease of the impact of the destruction reactions is less pronounced than in the C17.2-cell-like system. Even if APC is strongly mutated (mutant m10), the impact of the destruction reactions in the hepatocyte-like model is almost as strong as under wild-type conditions (compare row i and row iv in Figure 2.47C). In the hepatocyte-like model, the feedback loop is strong. This causes a higher concentration of Axin than in the extract or the C17.2-cell-like model. Although the dissociation of the APC-containing complexes is increased by the mutations, the higher Axin concentration drives the formation of the destruction complex. Consequently, the inhibition of the destruction reactions (reactions 4 to 9, 14 and 15) still has a high impact if the feedback loop is strong. The feedback loop restores the sensitivity of these reactions.

Thus, in the hepatocyte-like system an inhibition affecting the destruction reactions has a strong impact on the β-catenin steady state under wild-type conditions and if APC is mutated. In contrast, in the extract model, an inhibition of the destruction reactions has only a weak effect on the unstimulated β-catenin steady state if APC is mutated. This also holds for the C17.2-cell-like system carrying a strong APC mutation. If the APC mutation is weak, the inhibition of a destruction reaction in the C17.2-cell-like model has a stronger effect than in the extract model. In summary, depending on the strength of the mutation, distinct inhibitions have different effects in the different analysed models. An inhibition working efficiently in the hepatocyte-like model might have little or no effects in the C17.2-cell-like or the extract model. An inhibition that has no or a little effect in the extract model if APC is mutated might have a strong effect in the cellular models under the same mutated conditions.

2.4.8   Discussion

In the introduction (section 1) it is mentioned that a detailed mathematical model can be used versatilely. This includes the reflection of experimental data, the generation of hypotheses to explain experimental observations, the suggestion of experiments to distinguish between possible mechanisms and the prediction of the effects of mutations or drugs. In section 2.4, various of these aspects are used to get insights into cell-type-specific properties of Wnt/β-catenin signalling, effects of the feedback mechanism via Axin-2 and the binding of β-catenin and E-cadherin. Effects on the β-catenin steady states as well as on the temporal behaviour are of interest.

Experimental data

The dynamics of β-catenin protein and the mRNA of Axin-2 have been measured in primary hepatocytes and C17.2 cells upon pathway activation by the same kind of stimulus. This allows the comparison of the dynamics in the two cell-types. Differences in the unstimulated steady state concentration, the increasing phase and the decreasing phase of β-catenin have been detected. Furthermore, the relative concentration changes differ. Concerning the absolute protein levels one should mention that they have been measured in ng/mg whole cell protein. Hence, the levels are not the concentration per cell. However, for the simulations it is assumed that a higher value in the level in ng/mg whole cell protein implies a higher level in the concentration per cell.

To our knowledge, a comparison of time course data of β-catenin protein and the mRNA of Axin-2 upon pathway activation in two different cell-types has not been published before. In an earlier publication, β-catenin and Axin-2 dynamics have been presented for MDA 231 cells, which are breast carcinoma cells (Lustig et al., 2002). In this publication it has been focussed only on one cell-type. Furthermore, *in vivo* data as well as β-catenin dynamics and target gene activity in one cell-type (HEK293 cells) upon pathway activation with different stimuli has been published (Naik and Piwnica-Worms, 2007).

The experimental data of two cell-types can be described by the same pathway structure

With the cellular model one is able to reproduce qualitatively the temporal behaviour of β-catenin observed in murine primary hepatocytes and C17.2 cells, Figure 2.37. We found that the experimentally observed differences in the unstimulated β-catenin steady state as well as the concentration range, the increasing phase and the decreasing phase of β-catenin can be explained qualitatively by the cellular model. For both cell-types, the network structure is the same. To describe the differences experimentally observed in the two cell-types it is sufficient to assume distinct total APC and E-cadherin levels as well as a different strength of the feedback via Axin-2. These differences are based on experimental data (Figure 2.28 and Table 1). Nevertheless, one cannot exclude that additional differences in the concentrations of regulators or proteins involved directly in the pathway contribute to the different dynamics as well.

The decreasing phase observed in primary hepatocytes stimulated with the GSK3 inhibitor SB216763 could not be reproduced exactly by the mathematical model. Two hypotheses about the underlying reasons are stated. They are theoretically investigated, resulting in the prediction that experiments with stimulations at two successive time points would allow to distinguish between them (Figure 2.38). The corresponding experiments suggested by mathematical modelling indicate that the decrease in the β-catenin concentration is not only a consequence of the Axin-2 feedback nor determined exclusively by the degradation of the stimulus-mediating substance (Figure 2.39). To generate the observed dynamics, both mechanisms seem to work in combination. The involvement of further regulatory mechanisms cannot be excluded. It is possible that additional mechanisms contribute, such as direct regulatory mechanism like further feedback mechanisms or more indirect mechanisms via crosstalks with other pathways. Additional side effects by inhibiting the GSK3 are conceivable as well. SB216763 attacks the kinase GSK3 which is also involved in various other signalling pathways and in metabolic events (Doble and Woodgett, 2003).

Discussion of the effects of the single model extensions

The extract model of Wnt/β-catenin signalling is extended by two features that are important to cover additional cellular properties: i) the binding of β-catenin to E-cadherin that reflects the involvement of β-catenin in cell-cell adhesion, and ii) the transcriptional feedback loop via Axin-2.

Integrating the binding of β-catenin and E-cadherin does not affect the steady state of unbound β-catenin but the temporal behaviour (Figure 2.30, Figure 2.35). E-cadherin acts like a buffer. The fact that the steady state is not affected by E-cadherin is due to the assumption that the E-cadherin concentration stays constant over time. We applied the assumption of Lee and colleagues (Lee et al., 2003) of constant protein concentrations of Dsh, GSK3, APC and TCF also for E-cadherin since there is no indication of a fast E-cadherin turnover in the literature. If a turnover of E-cadherin would be allowed, E-cadherin would also have an effect on the steady state concentration of β-catenin. Such an E-cadherin turnover has been considered in the model developed by van Leeuwen and colleagues (van Leeuwen et al., 2007). Further experimental investigations of the role of E-cadherin are intended. An overexpression of E-cadherin or an E-cadherin knock-down will clarify whether or not E-cadherin has an effect on the β-catenin steady states. For prospective analyses, one may consider that Wnt/β-catenin signalling is discussed to affect the expression of E-cadherin negatively (Huber et al., 1996; Jamora et al., 2003; ten Berge et al., 2008). This would provide an additional regulatory loop within the signalling pathway. The transcriptional repression of E-cadherin by β-catenin has been included in the model of Shin and colleagues (Shin et al., 2010). However, they have not taken into consideration the binding of E-cadherin and β-catenin but have examined the expression of E-cadherin as readout of epithelial-mesenchymal transition.

Due to the feedback loop via Axin-2 the unstimulated steady state level of β-catenin is reduced compared to the extract model (Figure 2.31 and Figure 2.32). This is caused by the fact that the feedback loop is active even without pathway activation. The stronger the feedback loop, the lower the β-catenin concentration is. To verify this, experiments would be of interest in which the β-catenin concentration of one cell-type is detected if the strength of the Axin-2 feedback is altered. Unfortunately, to date it is impossible to perturb the Axin-2 feedback loop in experiments. In several publications it has been stressed that the effect of Axin-2 on the β-catenin level is very low in resting cells and only mediated in stimulated cells. This has been shown for breast carcinoma cells (MDA MB231) (Lustig et al., 2002). However, in resting cells, the Axin-2 protein has been detected, e.g. in mouse embryonic stem cells (Doble et al., 2007). Also the Axin-2 mRNA has been detected in primary hepatocytes and C17.2 cells (Figure 2.28) and in HEK293 cells (Choi et al., 2004). If the Axin-2 protein or the mRNA exists in the cell, one can assume that Axin-2 is available in the cell and might have an effect on the unstimulated β-catenin steady state. In the mathematical model, there is a considerable effect of Axin-2 on the unstimulated β-catenin steady state (Figure 2.31 and Figure 2.32). Possibly, the model should be adjusted such that the effect in unstimulated cells is lower, leading to a smaller decrease of the unstimulated steady state of β-catenin compared to the level in the extract system. The temporal behaviour of β-catenin is also affected by the Axin-2 feedback loop. It enables the decreasing phase (Figure 2.31) and affects the increasing phase (Figure 2.35).

With respect to the Axin-2 feedback one has to keep in mind that the Axin concentration is low is the extract model (Lee et al., 2003). This leads to simplifications of the model including the neglection of the Axin/GSK3 and Axin/β-catenin complexes. Due to the included feedback loop, the Axin concentration is increased. If the feedback loop is very strong, the Axin concentration is highly increased. This may necessitate the abrogation of the simplifications and an analysis of the system including the additional Axin-containing complexes. The Axin-2 feedback loop has been integrated in different ways in other models (see also section 2.1.2). Cho and colleagues (Cho et al., 2006) have included the feedback by direct activation of the Axin production by β-catenin and β-catenin/TCF. Due to this straight connection the delay of the effect of the feedback loop on the β-catenin concentration has been neglected. As we are not only interested in the steady state but also in the influence on the temporal behaviour, this model extension is not appropriate for our analyses. However, partially their realisation of the feedback loop is adapted (see section 2.4.2). Wawra and colleagues have enclosed the Axin-2 feedback loop as well (Wawra et al., 2007). They have applied a more general approach by using delayed differential equations. In the model introduced by Goldbeter and Pourquié (Goldbeter and Pourquie, 2008), the feedback via Axin-2 is also enclosed. Their whole model differs from the extract model as they have focussed on the crosstalk to other signalling pathways using a minimal model of Wnt/β-catenin signalling.

Effects mediated by the alteration of the total APC concentration

The experiments suggest that the APC concentration is different in the different cell-types (section 2.4.1). APC has a strong impact on the steady state (Figure 2.30 and Figure 2.32) and the β-catenin dynamics (Figure 2.35). As already discussed in sections 2.2 and 2.3, APC interacts with β-catenin in two ways: i) in the destruction complex, facilitating the degradation of β-catenin, and ii) by the formation of the APC/β-catenin complex. In the case of low total APC concentrations, this complex causes a considerable decrease of the APC concentration diminishing the β-catenin degradation as less destruction complex is composed (positive feedback on the β-catenin concentration). However, this holds only for low APC concentrations and might be biologically irrelevant. Nevertheless, for prospective theoretical analyses it might be of interest to analyse the different effects resulting from the coexistence of a positive feedback (via the APC/β-catenin complex) and a negative feedback (via Axin-2). Further experimental investigations on the function of APC are intended. Varying APC concentrations, either increased concentration by overexpression or decreased concentration resulting from application of siRNA against APC, will provide a deeper insight in the effects mediated by APC. Modelling results suggest that the β-catenin concentration is increased with decreasing APC concentrations. Furthermore, the simulations lead to the prediction that t_{max} is increased if the APC concentration is strongly increased.

Cell-type specificity and effects of mutations

Cell-type-specific responses to stimuli are of high importance. Revealing the reasons of different responses would for instance contribute to the understanding of the development of tissue-specific properties and tissue-specific diseases. Furthermore, it would make a contribution to the development of drugs with high target specificity.

In section 2.4.7, the effects of APC mutations of different strength are analysed theoretically. They are also object of research in section 2.2.3. Section 2.4.7 focuses on their impact on the β-catenin steady state and its temporal behaviour in the hepatocyte-like and the C17.2-cell-like model. The absolute and relative concentration changes are regarded. So far, there is no experimental time course data available, investigating the effects of APC mutations in different cell-types. Analyses have focussed on time course data in one cell-type (Lustig et al., 2002) or the β-catenin steady states or the reporter gene expression in various cell-types (Kohler et al., 2010; Kohler et al., 2009; Kohler et al., 2008; Yang et al., 2006).

The theoretical analyses reveal that with respect to the relative temporal behaviour of β-catenin, the APC mutations m1 and m5 do only have a small effect in the hepatocyte-like model but affect the relative β-catenin dynamics strongly in the C17.2-cell-like model (Figure 2.46). Due to the strong feedback loop in the hepatocyte-like model, the relative effect of these APC mutations is almost compensated. The changes in the relative β-catenin time courses are small. If the feedback loop is less strong like in the C17.2-cell-like model, the effects of the APC mutations are only weakly compensated. The relative temporal β-catenin behaviour is affected if APC is mutated. Analysing the absolute temporal behaviour of β-catenin shows that an APC mutation causes a higher β-catenin concentration under all analysed conditions. The stronger the mutation, the higher the unstimulated β-catenin steady state is (see also section 2.2.3). This is in accordance with experimental findings (Kohler et al., 2009). Time resolved experiments using cells with distinct APC mutations and feedback loops of different strengths would be of high interest to verify the theoretical findings.

Depending on the cell-type in which the mutation occurs, specific mutations affect the relative temporal behaviour of β-catenin differently. If one assumes that the relative concentration change is the relevant pathway readout as provided by the findings of Goentoro and colleague (Goentoro and Kirschner, 2009), this indicates that mutations establish distinct consequences in distinct cell-types or tissues. For instance, the APC mutation m5 affects the C17.2-cell-like system strongly while this APC mutation has only a little effect in the hepatocyte-like system. Those differences might be compared to the observation that different mutations are overrepresented in different types of tumours and are rarely detected in other types of tumours. In colon carcinoma, up to 85% of the tumours show a mutation in APC (Giles et al., 2003). In contrast, in hepatocellular carcinoma, an APC mutation is rare (Villanueva et al., 2007). Following our line of argument, an APC mutation of intermediate strength does not affect the relative β-catenin behaviour in the hepatocyte-like model. Consequently, compared to the wild-type case, no changes in the fold-change-dependent gene expression are induced, and thus, this APC mutation is not related to the development of HCC. Additional analyses of other cell-type-specific models and their comparison with experimental observation about the occurrence of specific mutations in specific tissues are needed for further verification. However, that could be an indicator that also in mutated cases the β-catenin fold-change is the relevant pathway outcome as considered for wild-type conditions (Goentoro and Kirschner, 2009).

For ERK signalling, the temporal behaviour of phosphorylated ERK (ERK^{PP}) has been analysed experimentally and theoretically in two cell-types that developed feedback loops of different strength (Kiel and Serrano, 2009). In the analyses, Kiel and Serrano have found that mutations within the feedback loop have different effects on ERK^{PP} depending on whether the feedback loop is strong or weak. If the feedback loop is strong, the effects of mutations are diminished. In contrast, if the feedback loop is weak, the mutations strongly affect the ERK^{PP}

dynamics. As the mutations do not affect the unstimulated steady state they have been able to analyse the absolute temporal behaviour of ERK^{PP}. In the Wnt/β-catenin pathway, the APC mutations lead to an increased unstimulated steady state concentration of β-catenin. Therefore, the effects of mutations cannot be compared as easy as in the ERK pathway. However, the relative β-catenin changes show a similar feature as in the case of ERK^{PP} in the ERK pathway. If the feedback loop is strong, the effect of the APC mutation is weak. If the feedback loop is weak, the APC mutation has a strong effect on the relative β-catenin time course. Regarding the relative β-catenin concentration, the feedback loop makes the Wnt/β-catenin pathway more robust against mutations in APC. This is comparable with the effects analysed in the ERK signalling pathway.

Additionally, drugs may work differently in distinct cell-types carrying APC mutations of different strength. Regarding the steady state it is shown by the modelling approach that inhibitions that affect the steady state strongly under wild-type conditions can just have little effects if APC is strongly mutated. This also depends on the kind of model; inhibitions that do not affect the steady state in the extract model or the C17.2-cell-like model can affect the steady state in the hepatocyte-like model. The inhibitions are also investigated with respect to their effects on the β-catenin fold-change (Figure C.2, Appendix C.2). Concerning the fold-change it holds that inhibitions that have a strong effect in the wild-type model might have weaker or stronger effects if APC is mutated. Whether the effect is enhanced or diminished depends on the model (extract, C17.2-cell-like or hepatocyte-like model) and the strength of the APC mutation. The analysis also reveals that for particular reactions such as the β-catenin production, the type of effect depends on the type of model and the strength of the APC mutation. This is in contrast to the effects on the β-catenin steady state which differ in the strength but always remain the same type of effect. Thus, to find an effective drug against the consequences of a specific mutation one has to consider the type of mutation and the properties of the specific cell-type.

Outlook

Further mutations

In this thesis, the effects of APC mutations of different strength are investigated. Depending on the kind of the cellular model, the mutations affect the β-catenin steady state and the dynamics differently. Wolf and colleagues (Wolf et al., 2007) have presented in their work on ERK signalling that mutations at different levels of the pathway have distinct consequences on the pathway readout and could be more or less effective. This also depends on the regulatory mechanisms within the system. With respect to ERK signalling it was also shown experimentally that the effects of inhibitors vary depending on the kind of mutation (Cichowski and Janne, 2010).

For Wnt/β-catenin signalling, the theoretical investigation of additional APC mutations is of interest (already discussed in section 2.2). Moreover, it will be of particular interest to investigate mutations of other pathway components in the cellular model. It can be asked which type of mutations will have stronger effects in the hepatocyte-like than in the C17.2-cell-like model and vice versa. Are there mutations affecting both cellular systems similarly? In hepatocellular carcinoma, APC mutations are rare. We speculated that one reason might be that in the hepatocyte-like model APC mutations do not have a strong effect on the β-catenin fold-change. In HCCs, β-catenin is frequently mutated and also mutations in Axin-1 and Axin-2 have been detected (Taniguchi et al., 2002), see also section 2.3. In some tumours, both Axin-1/2 and β-catenin are mutated. We state the question what the effects of β-catenin or Axin-2 mutations will be. To analyse the effects of the β-catenin mutations that cannot be phosphorylated bound to the destruction complex, the cellular model will be combined with the mutated model of Wnt/β-catenin signalling which is analysed in section 2.3. This allows the investigation of the effects of mutated β-catenin in different cellular systems as well as the analysis of effects of two or even more mutations coexisting in one cell.

3 Modelling the non-canonical NF-κB pathway

NF-κB (nuclear factor κ-light-chain-enhancer of activated B cells) signalling is involved in inflammatory responses and adaptive immunity and is linked to several diseases. In section 3.1, a biological overview of the NF-κB signalling is provided with a focus on the non-canonical NF-κB signalling branch. A mathematical model for this branch is developed in section 3.2. It is based on literature, discussions with experimental partners of Claus Scheidereit's group at the Max-Delbrück-Center for Molecular Medicine, and experiments performed by Z. Buket Yilmaz, a member of Claus Scheidereit's laboratory. The model is analysed and permits to make predictions that will be verified experimentally. With this combined theoretical and experimental approach, we aim to understand the non-canonical NF-κB signalling pathway and its long-term signalling. As non-canonical NF-κB signalling is often altered in diseases, modelling will help to understand the dysregulation and could predict effective intervention strategies.

3.1 Introduction in NF-κB signalling

NF-κB signalling is involved in inflammatory responses, innate and adaptive immunity, cell proliferation and cell death. Its dysregulation is linked to inflammatory, neoplastic and cardiovascular diseases as well as to cancer (Baldwin, 2001; Basseres and Baldwin, 2006; Courtois and Gilmore, 2006; Dejardin, 2006; Karin, 2006; Kumar et al., 2004; Staudt, 2010).

NF-κB signalling is divided into two branches, the canonical (classical) and the non-canonical (alternative) NF-κB signalling pathway (Bonizzi and Karin, 2004). These two branches operate on different time scales. The canonical pathway reacts fast on a stimulus being active within minutes. In contrast, the activation of the non-canonical branch is much slower requiring several hours (Sun, 2011). Both pathways use specific pathway components but also share some proteins.

Several proteins belong to the family of NF-κB that form different homo- and heterodimers: Rel-proteins (RelA (p65), RelB, c-Rel), the precursor p105 (NF-κB1) and its cleavage product p50 as well as the precursor p100 (NF-κB2) and its cleavage product p52. The respective precursor proteins are ubiquitinated, and subsequently, their C-terminal regions are proteolytically degraded, resulting in the active DNA binding forms p50 and p52. The p50/RelA heterodimer represents the complex that is regulated by the canonical NF-κB pathway. However, other NF-κB complexes exist as well. Non-canonical NF-κB signalling controls the concentration of p52/RelB.

Here, a short overview of the canonical NF-κB pathway is given. Non-canonical NF-κB signalling is described in more detail. Further information about NF-κB signalling could be found under www.nf-kb.org.

Until now, mathematical models have been focussed on the canonical NF-κB branch (Ashall et al., 2009; Basak et al., 2007; Cheong et al., 2008; Hoffmann et al., 2002; Kearns et al., 2006; Lipniacki et al., 2004; Mathes et al., 2008; Shih et al., 2009; Tay et al., 2010) which have been reviewed recently by Cheong and colleagues (Cheong et al., 2008). Recent modelling approaches have also taken into account a crosstalk of canonical and non-canonical NF-κB signalling (Basak and Hoffmann, 2008; Basak et al., 2007; Shih et al., 2011). However, a detailed model of the non-canonical branch is not yet published.

Biological background of canonical NF-κB signalling

Canonical NF-κB signalling is involved in the regulation of inflammatory responses (Karin and Staudt, 2010). Primarily, canonical NF-κB signalling is mediated by the heterodimer p50/RelA. The precursor protein p105 is constitutively processed to p50 that binds RelA. The p50/RelA complex is transcriptionally active. In resting cells, p50/RelA is inhibited by IκB (inhibitor of κB binding) molecules, which form a complex with p50/RelA. The molecules IκBα, IκBβ, IκBγ, IκBε, and Bcl3 belong to the family of IκB proteins. They have different affinities to different NF-κB complexes. Binding of IκBs retains p50/RelA in the cytoplasm. A stimulus induces the degradation of the inhibitor and thereby the release of p50/RelA. The canonical branch is activated by various stimuli including cytokines, viruses, genotoxic agents or oxidative stress. One of the best studied ligands is TNFα (tumour necrosis factor). The stimulus is transduced from the cell-surface receptor via several proteins such as TRAF (TNF receptor-associated factor) proteins and RIP (receptor interacting protein) to the kinase IKKβ (IκB kinase β). IKKβ is part of the IKK complex (IκB kinase complex) containing the two catalytic kinases IKKα and IKKβ and the scaffold protein NEMO (NF-κB essential modifier; also called IKKγ). For the canonical NF-κB signalling branch, IKKβ and NEMO are essential but not IKKα. IKKβ is activated by phosphorylation in the activation loop (T loop) enabling the kinase to phosphorylate IκBs. The IKKβ-mediated phosphorylation marks IκB for ubiquitination and subsequent proteasomal degradation. This results in the release of p50/RelA followed by its nuclear translocation and binding to the specific DNA sites (κB sites) to regulate the expression of various target genes. Among others, target genes encode for proteins involved in canonical or non-canonical NF-κB signalling, such as IκBα, TRAF1, TRAF2 and p100 (Hayden and Ghosh, 2004; Perkins, 2007).

Biological background of non-canonical NF-κB signalling

Non-canonical NF-κB signalling is critically involved in the development of a functional immune system by contributing to the development of primary (thymus) and secondary lymphoid organs (lymph nodes and Peyer's patches) as well as in B cell development (Scheidereit, 2006; Sun, 2011). A dysregulation of this pathway is linked to inflammatory, neoplastic and cardiovascular diseases as well as to cancer (Dejardin, 2006; Staudt, 2010).

This signalling branch leads to the processing of the precursor protein p100 (Xiao et al., 2001). The resulting p52-containing complexes (primarily p52/RelB) are transcriptionally active, regulating the expression of various target genes including chemokines such as B lymphocyte chemokine (BLC) (Dejardin et al., 2002). Several stimuli are able to activate the non-canonical signalling pathway by interacting with receptors of the TNF (tumour necrosis factor) receptor family including lymphotoxin β receptor (LTβ receptor, LTβR), B cell-activating factor receptor (BAFFR) and receptor activator of NF-κB (RANK)(Claudio et al., 2002; Dejardin et al., 2002). It should be noted that these stimuli do not activate the non-canonical branch exclusively, but also the canonical signalling which precedes the non-canonical activation (Scheidereit, 2006). Here, we concentrate on the activation of non-canonical NF-κB signalling via LTβR. A sketch of the non-canonical signalling pathway is shown in Figure 3.1.

NIK (NF-κB inducing kinase) plays a central role in the non-canonical NF-κB signalling pathway (Thu and Richmond, 2010). Its concentration is strictly regulated. The exact mechanism is still not fully understood and is under intensive experimental investigation (Razani et al., 2010; Vallabhapurapu et al., 2008; Zarnegar et al., 2008). Several proteins are involved in this regulation such as TRAF2, TRAF3 and cIAP1 (cellular inhibitor of apoptosis protein 1) or cIAP2 (hereafter referred to as cIAP1/2). Recent findings have led to the suggestion that TRAF2 and cIAP1/2 form a pre-complex. An additional pre-complex is formed by TRAF3 and NIK. By the association of TRAF2 and TRAF3, the two pre-complexes form a combined complex (Zarnegar et al., 2008) which is referred to as the destruction complex in the following. Within the destruction complex, cIAP1/2 is brought into close adjacence to NIK, enabling cIAP1/2 to ubiquitinate NIK. This ubiquitination leads to the degradation of NIK (Zarnegar et al., 2008). Due to the continuous degradation of NIK, the signal is not transduced further.

A stimulus interrupts the continuous degradation of NIK by disturbing the destruction complex. The stimulation leads to the multimerisation of the LTβ receptor enabling the receptor to transduce the signal to intracellular components including TRAF2 and TRAF3 (Sanjo et al., 2010). This results in an ubiquitination-cascade that leads to the ubiquitination of TRAF2 and TRAF3 by cIAP1/2 and their subsequent proteasome-dependent degradation (Vallabhapurapu et al., 2008). Consequently, the destruction complex disassembles (Vallabhapurapu et al., 2008; Zarnegar et al., 2008), leading to NIK accumulation and signal transduction. Therefore, NIK is phosphorylated (Lin et al., 1998). Until now, the regulation of this modification is not completely understood. Phosphorylated NIK is able to phosphorylate the kinase IKKα (IκB kinase α) in its activation loop (Ling et al., 1998; Senftleben et al., 2001). While in the canonical branch IKKβ and NEMO are required, the signalling of the non-canonical branch depends on IKKα (Senftleben et al., 2001). Its phosphorylation enables IKKα to phosphorylate i) the precursor p100, leading to the transmission of the signal, and ii) NIK, resulting in the destabilisation of NIK (Razani et al., 2010). The latter acts as an additional negative regulatory mechanism to control the concentration of NIK. The precursor p100 is phosphorylated at N- and C-terminal serine residues (Ser[99], Ser[108], Ser[115], Ser[123], and Ser[872]). For this step, NIK seems to be necessary as well (Xiao et al., 2004). Another study has suggested the involvement of p100 SUMOylation (SUMO stands for small ubiquitin-like modifier) in the regulation of p100 phosphorylation (Vatsyayan et al., 2008). As a consequence of the p100 phosphorylation, β-TrCP is recruited to the precursor and mediates the polyubiquitination of p100 at the lysine residue Lys[855] (Amir et al., 2004). This results in its C-terminal degradation, generating the cleavage product p52. Additionally, a constitutive processing of p100 has been assumed that is independent of NIK (Qing and Xiao, 2005). The generation of p52 results in the formation of p52/RelB heterodimers which enter the nucleus and regulate the transcription of various target genes (Bonizzi et al., 2004; Dejardin et al., 2002). Some of the target genes encode for members of the non-canonical pathway, such as RelB and TRAF3. In addition, the expression of components of the canonical NF-κB pathway is regulated, e.g. IκBα.

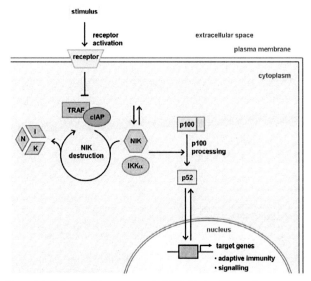

Figure 3.1: Scheme of the non-canonical NF-κB signalling pathway.

Without a stimulus, NIK (NF-κB inducing kinase) is degraded via a mechanism involving TRAF2, TRAF3 (TRAF) and cIAP1/2 (cIAP; cellular inhibitor of apoptosis protein). Upon receptor activation, the degradation is inhibited. Accumulation of NIK leads to IKKα (IκB kinase α) activation, enabling this kinase to phosphorylate p100. The phosphorylation of p100 is the prerequisite for its processing to its cleavage product p52. Complexes containing p52 enter the nucleus and regulate the expression of specific target genes.

3.2 Development and analyses of a mathematical model of the non-canonical NF-κB signalling pathway

3.2.1   The mathematical model

In this work, a first model of the non-canonical NF-κB signalling pathway is formulated. It is based on the detailed information provided in the biological literature and by the expertise of our experimental partners Claus Scheidereit and Z. Buket Yilmaz at the Max-Delbrück-Centrum for Molecular Medicine. The model includes the core proteins TRAF3, NIK, IKKα, p100, p52 and RelB.

The model is developed to understand the mechanisms and the effects of long-term signalling. It concentrates on the non-canonical pathway and neglects crossregulatory features with the canonical signalling branch. The canonical branch acts much faster than the non-canonical branch. Therefore, a separate investigation of both branches is reasonable. The developed model of the non-canonical NF-κB signalling pathway covers the effects of protein-protein interactions and modifications, mainly in the cytoplasm. For two components, a nucleo-cytoplasmic shuttle is included. The nuclear components are superscripted with "nuc" in the following. The volume differences between cytoplasm and nucleus are not included. Regarding the modifications, single and multiple modifications are not distinguished from each other. In any case, the modifications are considered as one step processes. In cases of phosphorylation or ubiquitination, the modified components are superscripted with "PP" or "Ub", respectively. So far, the model does not contain gene expression. Detailed information about the mechanisms regulating the gene expression is still missing. Hence, the transcriptional regulator p52/RelB^{nuc} is regarded as pathway readout.

The reaction scheme underlying the mathematical model is shown in Figure 3.2. The behaviour of the considered components and complexes is described by a set of 14 ordinary differential equations incorporating 28 reactions. It is given by the equations D.1 to D.45 (Appendix D). For the rate equations mass action kinetics are assumed. The rate constants are given in Table 2 in section 3.2.3.

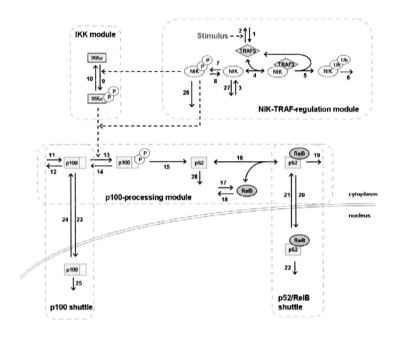

Figure 3.2: The reaction scheme of the non-canonical NF-κB pathway.
The model contains the NIK-TRAF-regulation module, the IKK module, the p100-processing module as well as the nucleo-cytoplasmic shuttle of p100 and p52/RelB. The individual modules are labelled and framed with dashed grey lines. See text for details of the single reaction steps. The phosphorylation of a component is indicated by "P", ubiquitination is indicated by "Ub". Numbers next to arrows denote the number of the particular reaction. One-headed arrows denote reactions taking place in the indicated direction. Dashed arrows represent activation steps and double-headed arrows illustrate reversible binding reactions.

The model contains the following processes and components: The concentration of NIK is regulated by its ubiquitin-dependent degradation. NIK is produced and degraded (reactions 3 and 27, respectively). It can reversibly bind TRAF3 (reaction 4) which is produced (reaction 1) and degraded (reaction 2) itself. A stimulus amplifies the TRAF3 degradation. Within the TRAF3/NIK complex, NIK is modified. Modified NIK (NIK^{Ub}) dissociates from TRAF3 (reaction 5) and is degraded (reaction 6). NIK that is not bound to TRAF3 is phosphorylated (reaction 7). Phosphorylated NIK (NIK^{PP}) is able to phosphorylate IKKα (reaction 9) resulting in the activation of the IKKα. In cooperation with NIK^{PP}, IKKα^{PP}

phosphorylates p100 (reaction 13). The phosphorylations of IKKα and NIK are reversible (reactions 10 and 8, respectively). Additionally, NIK^{PP} is degraded (reaction 26). For IKKα, a conservation relation is supposed. No synthesis or degradation is included.

The precursor p100 is produced and degraded (reactions 11 and 12, respectively). Phosphorylated cytoplasmic p100 ($p100^{PP}$) is processed to its cleavage product p52 (reaction 15). $p100^{PP}$ can also be dephosphorylated again (reaction 14). Cytoplasmic p52 reversibly binds RelB (reaction 16). In addition, a degradation of p52 is included (reaction 28). RelB is produced and degraded (reactions 17 and 18, respectively). The resulting p52/RelB-complex shuttles between the cytoplasm and nucleus (reactions 20 and 21). Nuclear p52/RelB ($p52/RelB^{nuc}$) is the complex that regulates the transcription of pathway-specific target genes and which is considered as the pathway readout. Cytoplasmic as well as nuclear p52/RelB are degraded (reactions 19 and 22, respectively). Like p52/RelB, p100 is able to shuttle between the cytoplasm and the nucleus (reactions 23 and 24). Nuclear p100 ($p100^{nuc}$) is degraded (reaction 25).

In the cytoplasm, three modules are depicted:

- the NIK-TRAF-regulation module,

- the IKK module, and

- the p100-processing module.

Furthermore, two modules exist that take into account the nucleo-cytoplasmic shuttles of p100 and p52/RelB.

Methods

The concentration of the system's components changes due to their production, degradation, modifications or binding to or release from a complex. They are described by a set of ODEs. The rates are functions of the substrates, reactants, effectors, and kinetic parameters.

The parameters are obtained by fitting the model to the time course data (section 3.2.2) using a Markov-Chain Monte Carlo (MCMC) approach (Press et al.), which is a random sampling method. Each state p_s is characterised by a set of parameters. A new state p_{s+1} depends only on the previous state p_s. A posterior probability distribution $P(p|d)$ of the parameters p given the data d is constructed:

$$P(p \mid d) = \frac{P(d \mid p) \cdot P(p)}{P(d)} .$$

By the term P(p), prior knowledge of parameters is included for instance that they have to be non-negative. The model is solved numerically to calculate P($d|p$). It is assumed that the error at each data point is Gaussian. Then P($d|p$) reads:

$$- \ln P(d \mid p) = \sum_i \frac{(x_i^{sim} - x_i^{data})^2}{2 \cdot \sigma_i^2}.$$

This is the residual sum where x_i^{data} is the ith measured data point with the standard deviation σ_i and x_i^{sim} is the calculated approximation of x_i^{data}. The experimental data shows a standard deviation between 0.05% and 60% (see section 3.2.2).

To sample the probability distribution, a Metropolis-Hastings algorithm is used (Press et al.). With this approach the parameter space is explored systematically. The algorithm uses a proposal density Q(p, p') depending on the current state p to generate a new proposed sample p'. To decide whether the new sample p' is accepted, the acceptance ratio α is calculated:

$$\alpha = \frac{P(p'|d) \cdot Q(p', p)}{P(p|d) \cdot Q(p, p')}.$$

If $\alpha > 1$, p' is accepted ($p' = p_{s+1}$). If $\alpha < 1$, p' will be accepted if α is greater than a which is a random number between 0 and 1. If $\alpha < 1$ and $\alpha < a$, the proposal p' is not accepted. In this case, the current state p is also used in the next iteration ($p = p_{s+1}$).

At the end, we obtain a parameter set for the specific model that is able to reproduce the available experimental data. Although time resolved data of numerous pathway components is used for the fitting, it does not allow an unambiguous fit. The parameters are underdetermined, resulting in different parameter sets that are able to describe the experimental data.

The model is fitted 20 times. Depending on the criteria, at least 4 parameter sets resulting from the fit are not in a biological realistic range. This is for instance indicated by the fact that they cause unrealistic concentrations of pathway components. These parameter sets are not further considered. However, there are still several non-unique parameter sets which describe the experimental data well. For the analysis of the model, one exemplary parameter set is chosen. In this parameter set, all parameter values are in a biologically realistic range (discussed in section 3.2.3). The values are given in Table 2. It is made certain, that the general observations and conclusions also hold if other fitted parameter sets, which are assessed to be reasonable, are used (not shown).

By the fitting approach a distribution of values for each parameter is obtained. If the residual sum converges, at least 150000 further steps have been simulated. We assess when the algorithm converges by plotting the residual sum for each parameter set. After an initial transient decreasing phase of the residual sum, it fluctuates around a mean value (plot not shown).

The median values of the distributions of the individual parameters are used for the simulations and calculation of the sensitivity coefficients. Please, note that although the residual sum has converged, the values of several parameters have not converged.

3.2.2   Experimental data and simulated time courses

Time course data of components of the non-canonical signalling branch has been generated. Z. Buket Yilmaz has performed three biological replicates of 12 hours time course experiments using mouse embryonic fibroblasts (MEFs). Samples have been taken every hour. The cells have been stimulated with agonistic AC.H6 antibody for LTβR stimulation or treated with HA4/8 antibody as control. The control data is not presented. The cells have been fractionated into the cytoplasmic and the nuclear fraction and analysed by Western blotting. The Western blots have been quantified using the freeware ImageJ that has been developed at the National Institutes of Health (http://rsb.info.nih.gov/ij/index.html). The resulting integrated densities of the band of the proteins of interest have been related to the integrated density of the corresponding loading control. For a better comparison, the time courses have been normalised. The data of cytoplasmic and nuclear p100 as well as the TRAF3 data are related to the concentration in the unstimulated cell. Cytoplasmic and nuclear data of p52 and RelB are related to the respective concentration reached at $t = 12$ h. The normalised quantified experimental time courses of the three independent experiments have been used to fit the model to the data. The experimental data is shown in Figure 3.3.

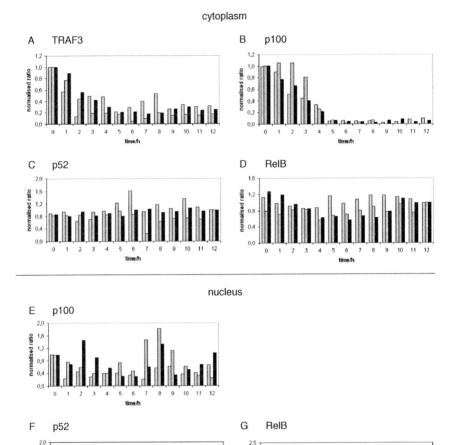

Figure 3.3: Relative temporal behaviour of cytoplasmic and nuclear components of the non-canonical NF-κB signalling pathway.

The relative concentration changes of the cytoplasmic components (A) TRAF3, (B) p100, (C) p52 and (D) RelB is shown as well as the concentration changes of the nuclear components (E) p100, (F) p52 and (G) RelB. MEFs have been stimulated with AC.H6 at $t = 0$ h. The relative concentration changes measured in three independent experiments are presented, experiment 1 (grey), experiment 2 (light blue) and (continued on page 148)

First, the temporal behaviour of the components in the cytoplasm is considered (Figure 3.3A to D). Due to the stimulus, the relative concentration of TRAF3 (Figure 3.3A) decreases within the first 4 hours. Afterwards, the concentration stays almost constant on a relative level of ~0.2 over the experimental time frame. The stimulus also induces a decline in the concentration of p100 as expected and published (Müller and Siebenlist, 2003; Yilmaz et al., 2003) (Figure 3.3B). After 5 hours, a very low relative concentration is achieved (~0.05). It stays low until 12 h. In contrast to the clear concentration changes of TRAF3 and p100, no clear concentration changes of cytoplasmic p52 (Figure 3.3C) and RelB (Figure 3.3D) have been detected.

The dynamics of the nuclear components is shown in Figure 3.3E to G. In resting cells, the concentration of nuclear p52 (Figure 3.3F) is very low. Before pathway stimulation, the relative p52 concentration is ~0.01. Due to the stimulation, the relative concentration of nuclear p52 increases as a result of the p100 processing. After 4 – 5 h, a maximum is reached. Its relative concentration is between 0.9 and 1.6. Thereafter, the relative concentration slightly decreases. The relative concentration changes of nuclear RelB (Figure 3.3G) and nuclear p100 (Figure 3.3E) do not show a clear trend. In these cases, the variation of the experimental data is high.

Continued legend of Figure 3.3.

experiment 3 (blue). The concentration changes of TRAF3, cytoplasmic p100, p52 and RelB as well as nuclear p100 are related to the respective concentrations of the unstimulated system. All other concentrations are related to the corresponding concentrations reached at 12 h.

Simulated time courses

The model of the non-canonical NF-κB pathway is fitted to the experimental data presented in Figure 3.3. In Figure 3.4, the time courses are shown that are simulated with the exemplary parameter set given in Table 2 (black lines). The dynamics of the pathway components are compared to the corresponding experimental data. The mean and standard deviation of the three experiments is calculated and presented in grey.

Corresponding to the experiments, the simulated temporal changes of the total concentrations of TRAF3, cytoplasmic p100, p52 and RelB, are shown. Total TRAF3 consists of unbound TRAF3 and the TRAF3/NIK complex. Total cytoplasmic p100 represents the sum of cytoplasmic unmodified p100 and phosphorylated p100. Total cytoplasmic p52 denotes cytoplasmic unbound p52 and the cytoplasmic p52/RelB complex, while total cytoplasmic RelB takes the cytoplasmic p52/RelB complex and unbound RelB into account.

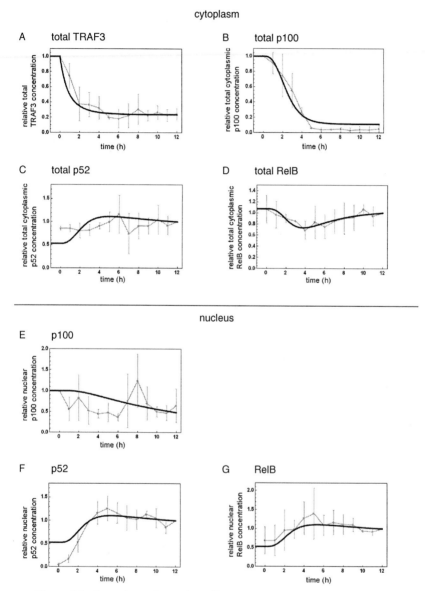

Figure 3.4: Comparison of experimentally measured and simulated time courses of components of the non-canonical NF-κB pathway.

The relative concentration changes of the cytoplasmic components (A) TRAF3, (B) p100, (C) p52 and (D) RelB are shown as well as the relative concentration (continued on page 151)

Figure 3.4 illustrates that the simulated time courses (black lines) fit well to the experimental data (grey lines).

The constant stimulus induces a decrease in the total TRAF3 concentration (Figure 3.4A). The concentration decreases within the first 4 hours, reaching a low relative concentration ~0.24. The relative concentration of total cytoplasmic p100 (Figure 3.4B) also decreases upon pathway activation. The decrease does not start immediately after stimulation but sets in approximately 30 min thereafter. The concentration decreases strongly for 6 h. A low relative concentration of approximately 0.1 is reached. Afterwards, the total p100 concentration stays on this low level. In the unstimulated model, the relative concentration of total cytoplasmic p52 is ~0.53 (Figure 3.4C). It increases within the first 5 h reaching a relative maximum of ~1.1 at ~5.3 h that is followed by a slight decrease. Induced by the stimulus, the relative total concentration of cytoplasmic RelB (Figure 3.4D) transiently decreases. The minimum of ~0.73 is reached after 3.8 hours.

In Figure 3.4E to G, the time courses of nuclear pathway components are shown. Within the 12 h time frame, the simulated concentration of nuclear p100 (Figure 3.4E) decreases by ~50%. The simulated dynamics of nuclear p52 and RelB (Figure 3.4F and G) are equal. In the model, both are represented by the nuclear p52/RelB complex. As the experimental data differs, the time courses of the two components are presented separately. In the unstimulated model, the relative concentration is approximately 0.53. The stimulus induces an increase of the concentration that sets in ~30 min after its application. The concentration increases until it reaches its relative maximum of approximately 1.1 at 5.3 hours. Between the time point at which the maximum is reached and 12 hours, the relative concentration slightly decreases.

Continued legend of Figure 3.4.

changes of the nuclear components (E) p100, (F) p52 and (G) RelB. For TRAF3, cytoplasmic p100, p52 and RelB, the changes of the total concentrations are shown. The concentration changes of total TRAF3, total cytoplasmic and nuclear p100 are related to the corresponding concentration of the unstimulated system. All other concentrations are related to the respective concentration reached at 12 h. The simulated dynamics are presented in black. The system is constantly stimulated at $t = 0$ h. For comparison, the experimentally measured time courses are additionally shown. The mean and standard deviation of the three experiments presented in Figure 3.3 are shown in grey.

The simulated time courses of cytoplasmic TRAF3 and p100 fit very well to the experimental data (compare black and grey lines in Figure 3.4A and Figure 3.4B). The general behaviour of nuclear p52 and RelB observed in experiments – the concentration increase until the maximum is reached at 5 h followed by a slight decrease – is reproduced in the simulations as well. However, compared to the measured concentration of nuclear p52 in resting cells, the simulated unstimulated steady state level of p52/RelB is too high. In contrast, the calculated unstimulated steady state of nuclear p52/RelB is lower than the level of nuclear RelB detected in experiments. These deviations result from the fact that the behaviour of nuclear p52/RelB is fitted to both the relative concentration change of nuclear p52 and nuclear RelB. The measured dynamics differ. While nuclear p52 increases there is no trend in the nuclear concentration change of RelB. One has to consider whether additional nuclear reactions such as the dissociation of the p52/RelB complex are reasonable. This would lead to a fit of the distinct entities. So far, a separate consideration of nuclear RelB and nuclear p52 is not included as there is less known about distinct functions of these two nuclear components. The other measured time courses do not show clear trends in the concentration changes over time. In part, the corresponding simulated dynamics show clear tendencies for instance the concentration change of total cytoplasmic p52 (Figure 3.4C). However, the comparison of the experimental and simulated dynamics reveals that the simulated time courses are in the range of the standard deviation of the experimental data. Especially the nuclear experimental data shows a high variance.

In addition to the measured time courses, the dynamics of additional selected pathway components are simulated. The relative dynamics of NIK^{PP}, IKKα^{PP}, p100^{PP} and unbound cytoplasmic p52 are shown in Figure 3.5.

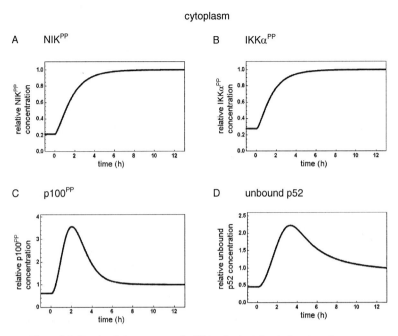

Figure 3.5: Simulated time courses of additional selected cytoplasmic pathway components of the non-canonical NF-κB activation.
The relative concentration changes of the cytoplasmic components (A) NIK^{PP}, (B) IKKα^{PP}, (C) p100^{PP} and (D) unbound p52 are shown. For all four relative time courses, the concentrations are related to the corresponding concentrations reached at $t = 12$ h. The model is constantly activated at $t = 0$ h.

The simulations presented in Figure 3.5 show that the relative dynamics of NIK^{PP} (Figure 3.5A) and IKKα^{PP} (Figure 3.5B) are very similar. Due to the stimulation of the pathway, the relative concentrations of both components increase until reaching the stimulated steady state levels. The ratio between stimulated and unstimulated steady state slightly differs for the two components. While for IKKα^{PP} this ratio is approximately 3.6, the ratio concerning NIK^{PP} is approximately 4.6. Upon pathway activation, the relative concentration of p100^{PP} increases and reaches its maximum at 2 h (Figure 3.5C). The relative concentration increases by a factor of ~6, followed by a decrease of the relative concentration. The main concentration change occurs until 6 h. The relative concentration of unbound p52 also increases if the pathway is activated (Figure 3.5D). The concentration increases by a factor of ~4.8, reaching the maximum at 3.3 h. Thereafter, the relative p52 concentration decreases. At $t = 12$ h, the stimulated steady state is not yet reached. Compared to the increase of the $p100^{PP}$ concentration, the increase in the p52 concentration is slower. The p52 maximum is reached approximately 1.3 h later than the $p100^{PP}$ maximum. The ratio between p52 maximum and unstimulated steady state concentration is slightly smaller than the ratio calculated for the $p100^{PP}$ dynamics.

Taken together, the simulated relative concentrations of NIK^{PP} and $IKK\alpha^{PP}$ increase until they reach their stimulated steady state levels. In contrast, the concentrations of $p100^{PP}$ and unbound p52 show an overshoot before reaching the steady state concentration.

3.2.3   Fitted parameters

So far, no parameter set for the non-canonical NF-κB signalling pathway has been published. In Table 2, the exemplary parameters are provided that are determined by fitting the model to the experimental data and that are used for all presented theoretical analyses.

Table 2: Parameters fitted for the mathematical model of non-canonical NF-κB signalling.
For each parameter the fitted median value is given. The values of the stimulus (k_2^{stim}) and the degradation rate of NIK^{Ub} are fixed and not fitted (marked by *). The total concentration of IKKα obeys the conservation relation. A total concentration of 5 nM is considered.
The numbers next to arrows in the model scheme in Figure 3.2 denote the number of the particular reactions. Nuclear components are superscripted with "nuc". Phosphorylated and ubiquitinated components are superscripted with "PP" and "Ub", respectively Components in a complex are separated by a slash.

Parameter	Short description	Value	
v_1	TRAF3 synthesis	0.0907 nmol·h^{-1}	
k_2	constitutive TRAF3 degradation	0.166 h^{-1}	
k_2^{stim}	inducible TRAF3 degradation	1.4 h^{-1}	*
v_3	NIK synthesis	8.468 nmol·h^{-1}	
k_4	formation of NIK/TRAF3 complex	153.141 nmol^{-1}·h^{-1}	
k_{-4}	dissociation of NIK/TRAF3 complex	1.191 h^{-1}	
k_5	release of NIK^{Ub}	26.198 h^{-1}	
k_6	degradation of NIK^{Ub}	50 h^{-1}	*
k_7	NIK phosphorylation	3.049 h^{-1}	
k_8	NIK^{PP} dephosphorylation	10.981 h^{-1}	

(Continued on page 155)

Continued Table 2

Parameter	Short description	Value
k_9	IKKα phosphorylation	69.269 nmol^{-1}·h^{-1}
k_{10}	IKKα^{PP} dephosphorylation	14.198 h^{-1}
v_{11}	p100 synthesis	3.202 nmol·h^{-1}
k_{12}	p100 degradation	0.0321 h^{-1}
k_{13}	p100 phosphorylation	11.943 nmol^{-2}·h^{-1}
k_{14}	p100^{PP} dephosphorylation	0.937 h^{-1}
k_{15}	p100^{PP} processing	6.503 h^{-1}
k_{16}	formation of p52/RelB complex	31.311 nmol^{-1}·h^{-1}
k_{-16}	dissociation of p52/RelB complex	1.333 h^{-1}
v_{17}	RelB synthesis	10.386 nmol·h^{-1}
k_{18}	RelB degradation	0.284 h^{-1}
k_{19}	p52/RelB degradation	0.272 h^{-1}
k_{20}	p52/RelB shuttle into the nucleus	14.909 h^{-1}
k_{21}	p52/RelB^{nuc} shuttle out of the nucleus	3.817 h^{-1}
k_{22}	p52/RelB^{nuc} degradation	0.0584 h^{-1}
k_{23}	p100 shuttle into the nucleus	0.0194 h^{-1}
k_{24}	p100^{nuc} shuttle out of the nucleus	0.0259 h^{-1}
k_{25}	p100^{nuc} degradation	0.063 h^{-1}
k_{26}	NIK^{PP} degradation	5.338 h^{-1}
k_{27}	NIK degradation	10.154 h^{-1}
k_{28}	p52 degradation	0.0143 h^{-1}

The parameters that are given in Table 2 are in a biologically realistic range. Additionally, the resulting absolute concentrations of pathway components take realistic values (not shown). Until now, kinetic parameters of reactions of the non-canonical NF-κB signalling branch have not been published. However, published experimental data permits to conclude about temporal characteristics of processes. In particular, these observations focus on the stability of proteins. The independent experimental observations are discussed with respect to the fitted parameters. The low degradation rate of p100 (k_{12}) indicates that unmodified cytoplasmic p100 has a long life span. Hence, it is a stable protein. This is in accordance with experimental observations (Müller and Siebenlist, 2003). In experiments in which the biosynthesis has been blocked, the p100 level has still been high after 8 hours. This is an indicator, that the degradation rate of p100 is low. The same holds for its cleavage product p52. Its degradation rate (k_{28}) is even lower. This fits to unpublished experimental data that indicates that p52 has a longer life span than p100 (personal communication Z. B. Yilmaz). The fitted degradation rate of RelB (k_{18}) indicates that RelB has a shorter life span than p52 and p100. However, it is still quite stable in comparison to other proteins. This finding is in agreement with experimental observations. Experiments with the biosynthesis inhibitor have revealed that 4 h after treatment, the RelB level has been decreased only slightly in comparison to the level in unstimulated cells. However, after 8 h, the RelB level has considerably been decreased (Leidner et al., 2008). In contrast to these stable proteins, the degradation rates of NIK (k_{27}) and NIK^{PP} (k_{26}) are high, indicating that NIK is an unstable protein in its phosphorylated and unphosphorylated form. This is in agreement with the experimental data. In resting cells, NIK is experimentally almost undetectable. NIK has been detected only if its degradation is inhibited by the application of proteasome inhibitors (Thu and Richmond, 2010). Besides conclusions on the stability of proteins, some further conclusions can be drawn. The parameter fitting shows that the nucleo-cytoplasmic shuttle of p52/RelB is very fast. This is also in agreement with experimental observations in which the occurrence of p52 has separately been measured in the cytoplasm and the nucleus (Coope et al., 2002). No delay between the occurrence in the cytoplasm and nucleus has been detected. This is an indication of a fast nucleo-cytoplasmic shuttle. The parameter fit also suggests that the formation of the NIK/TRAF3 complex (reaction 4) is a very fast process. It is known that TRAF3 binds NIK with high affinity (Zarnegar et al., 2008). Time-resolved data of this binding is not available. The phosphorylation of IKKα (reaction 9) is also a fast process. To my knowledge, time-resolved data of the activation process of IKKα has yet not been published.

3.2.4   Model predictions

In section 3.2.3, the implications of single parameters are discussed mainly related to the stability of the components of the pathway. In this section, the effects of the treatment with the protein biosynthesis inhibitor cycloheximide (CHX) are investigated. For the conditions of CHX treatment, the dynamics of selected pathway components are simulated (Figure 3.6). Furthermore, their temporal behaviour is shown if a combination of CHX treatment and the non-canonical stimulus is considered. In this context, CHX treatment is regarded as stimulus as well. For comparison, the dynamics resulting from application of the non-canonical stimulus are presented. The behaviour of total TRAF3, NIK^{PP}, total cytoplasmic p100 and nuclear p52/RelB are investigated. In the model, the cycloheximide treatment of cells is simulated by the inhibition of all protein biosynthesis reactions by 80%. The affected reactions are the syntheses of TRAF3 (reaction 1), NIK (reaction 3), p100 (reaction 11) and RelB (reaction 17).

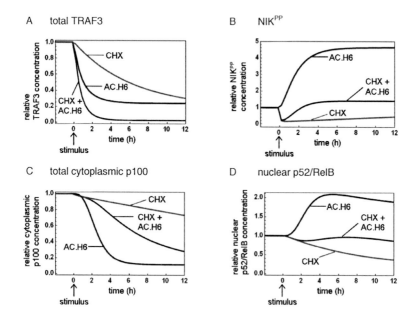

Figure 3.6: Predicted temporal behaviour of the selected pathway components upon application of different stimuli.

The simulated relative time courses of (A) total TRAF3, (B) NIK^{PP}, (C) total cytoplasmic p100 and (D) nuclear p52/RelB is shown. At $t = 0$ h, the system is stimulated by the non-canonical stimulus AC.H6 (black), the protein biosynthesis inhibitor cycloheximide (red, CHX), or both stimuli (blue, CHX + AC.H6). The application of cycloheximide is simulated as decrease of the all synthesis rates (reactions 1, 3, 11 and 17) by 80%.

Figure 3.6 shows that the three stimuli differently affect the temporal behaviour of total TRAF3, NIK^{PP}, total cytoplasmic p100, and nuclear p52/RelB. Cycloheximide administration leads to a decrease in the total TRAF3 concentration as less TRAF3 is produced. This decrease is not as fast as the decrease induced by the non-canonical stimulus (AC.H6) (compare red and black line, Figure 3.6A). However, both treatments cause a decrease of the total TRAF3 concentration. Hence, the effects are of the same kind. If the system is treated with both stimuli at the same time (blue line), the relative total TRAF3 concentration decreases even faster and reaches a lower relative concentration compared to the behaviour induced by the single stimuli. This is due to the fact that both treatments cause effects of the same type which act jointly. With regard to NIK^{PP}, the administration of AC.H6 or CHX mediates different kinds of effects. While the concentration of NIK^{PP} increases due to non-canonical stimulation (black line, Figure 3.6B), the concentration decreases if the biosynthesis is inhibited (red line). The strong decline in the latter is caused by the fact that less NIK is produced and rapidly degraded (via reaction 27). NIK also binds TRAF3 (reaction 4). Therefore, less NIK^{PP} is generated. The slight increase in the NIK^{PP} concentration that occurs over time originates from the decreased TRAF3 concentration if CHX is applied (red line Figure 3.6A). Since a lower amount of TRAF3 is available, less NIK binds TRAF3 and more NIK is phosphorylated. However, the concentration is very low. If the system is stimulated with AC.H6 and cycloheximide at the same time (blue line), the increase in the NIK^{PP} concentration at later time point is stronger pronounced, as the total TRAF3 concentration strongly decreases (blue line, Figure 3.6A). The total concentration of cytoplasmic p100 decreases in the presence of the non-canonical stimulus, upon CHX treatment and administration of both stimuli (Figure 3.6C). If the protein biosynthesis is inhibited, the total concentration of cytoplasmic p100 decreases slightly (red line). Only a low amount of p100 is degraded via the degradation reaction of p100 (reaction 12), and thus, the concentration stays quite high. This finding corresponds to the experimental observations that the p100 level does not change strongly upon cycloheximide treatment (Coope et al., 2002; Müller and Siebenlist, 2003) and is already discussed in section 3.2.3. The stimulation by a non-canonical stimulus leads to a considerable decline of the total concentration of p100 in the cytoplasm (black line) which is in agreement with the experimental data (Figure 3.3). The application of both substances causes a decrease of total cytoplasmic p100 that is neither as weak as in the case of cycloheximide treatment nor as strong as under the presence of the non-canonical stimulus. Hence, the administration of both substances leads to a lower decrease of p100 although the decrease of the TRAF3 concentration is very strong under these circumstances. Furthermore, if both stimuli are applied, the decline of total cytoplasmic p100 is delayed compared to the time course resulting from treatment with the non-canonical stimulus (compare blue and black line in Figure 3.6C). Both findings are caused by the altered availability of NIK^{PP}. Since less NIK^{PP} exists, the signal transduction to induce p100 processing is diminished. The delayed increase of the NIK^{PP} concentration is the reason for the delayed decrease of the p100 concentration. Experimental investigations have suggested that within the first 6 or 8 h, respectively, there is almost no decrease in the p100 concentration if a non-canonical stimulus is added in combination with CHX (Coope et al., 2002; Müller and Siebenlist, 2003). The simulations show a decrease which is not as strong as the decline caused by AC.H6 administration, showing the same general property. However, there is a considerable decrease of the p100 concentration and thus, a slight deviation between simulation and experiment. This might be diminished if a weaker efficiency of CHX would be assumed in the simulations. To my knowledge, the CHX efficiency is yet unknown. Last, the nuclear p52/RelB concentration is simulated under the three stimulation scenarios (Figure 3.6D). If the non-canonical stimulus is exclusively applied, the concentration of nuclear p52/RelB increases (black line in Figure 3.6D). In contrast, the treatment with the biosynthesis inhibitor leads to a decrease of that concentration since less precursor protein p100 and less RelB are produced (red line in Figure 3.6D). The concentration of total p52, including nuclear and cytoplasmic p52, also decreases upon CHX treatment (simulations not shown). In the experiments in which CHX is applied exclusively, the p52 expression has been shown to stay unchanged in lysates (Müller and Siebenlist, 2003) and whole cell extracts (Coope et al., 2002). This is a deviation between the modelling results and the experimental observations. Since the decrease of p52 is caused by the decrease of p100 and RelB this might indicate that p100 and RelB have even lower degradation rates as the rates arisen from the fitting. Additionally and again, the difference between the experimental observation and the simulation results would be reduced if a lower efficiency of CHX would be considered in the simulations. In the case of the treatment with both stimuli (blue line), the nuclear p52/RelB concentration changes only slightly. Initially, the concentration of $p52/RelB^{nuc}$ decreases slightly like in the case of CHX treatment. Afterwards, the concentration marginally increases. This is caused by the enhanced processing of p100.

The simulations predict that an exclusive treatment with cycloheximide or the application of cycloheximide in combination with a non-canonical stimulus cause different effects on the dynamics of different pathway components. The total TRAF3 concentration decreases the strongest if both stimuli are applied. The exclusive CHX treatment leads to a slighter decrease of the TRAF3 concentration than pathway activation with a non-canonical stimulus. The total concentration of cytoplasmic p100 also decreases slightly if the system is stimulated by CHX. In contrast to the behaviour of total TRAF3, the concentration of total cytoplasmic p100 decreases to a stronger extent by the pathway activation upon AC.H6 treatment than by the administration of both stimuli. The simulations further lead to the prediction, that the stimulation with both stimuli does not induce a strong concentration change of nuclear p52/RelB. In contrast, applying only CHX results in a decrease of the p52/RelB^{nuc} concentration while the non-canonical stimulus induces an increase of p52/RelB in the nucleus. These are the time courses that are experimentally distinguishable. Therefore, we suggest experiments with the three different stimuli and the detection of time course data for the verification of the model. Moreover, the data of such additional perturbation experiments can be integrated in the parameter fit and could contribute to an unambiguous determination of the model parameters.

3.2.5   Modelling mutations

In order to validate the model, it is examined whether the model is able to reproduce the behaviour experimentally observed in cases of mutations of pathway components. Only mutations of proteins included in the model are considered. In Table 3, the effects of selected mutations are provided. A summary of various mutations in the canonical and the non-canonical branch and their resulting phenotypes are given in the review by Gerondakis and colleagues (Gerondakis et al., 2006).

Table 3: Mutations of the proteins of the non-canonical NF-κB pathway.
The mutations and their effects observed in experiments are listed with the respective references. Furthermore, the realisation in the model and the simulated effects are given. The considered mutations affect TRAF3, NIK, IKKα and p100.
Simulation results that show a deviation to the experimental observations are coloured in red.
-/-: complete deletion of the respective gene, hence, a lack of the corresponding protein.
aly/aly: homozygotic alymphoplasia (aly) (Shinkura et al., 1999) mutation of the NIK gene. This mutation impedes the signal transduction from NIK to IKKα (Luftig et al., 2001).

Mutation	Experimental observation	Reference	Realisation in the model	Effects in the model
TRAF3$^{-/-}$	- increased NIK level - decreased p100 level - increased level of nuclear p52 - loss of inducibility of p100 processing by stimulation	(He et al., 2006) (Vallabhapurapu et al., 2008) (Sanjo et al., 2010)	$k_1 = 0$ nmol·h^{-1}	- NIK level is increased - p100 level is decreased - increased nuclear p52/RelB level - stimulation does not lead to any concentration change

(Continued on page 161)

Continued Table 3.

Mutation	Experimental observation	Reference	Realisation in the model	Effects in the model
TRAF3 without NIK interaction region	- decreased p100 level - increased p52 level	(Sanjo et al., 2010)	$k_4 = 0$ nmol^{-1}·h^{-1}	- decreased p100 level - increased total p52 level
NIK$^{aly/aly}$; NIK with inactive/with-out kinase domain	- defect in p100 ubiquitination - abolished p100 processing - defect in p52 production	(Xiao et al., 2001);	$k_9 = 0$ nmol^{-1}·h^{-1}	- high p100 level; no processing induced - no p52
NIK$^{aly/aly}$	- no p100 processing - defect in p52 production - stimulus-induced TRAF3 degradation is not affected (unpublished; Z. B. Yilmaz)	(Yilmaz et al., 2003)	$k_9 = 0$ nmol^{-1}·h^{-1}	- high total p100 level; no processing inducible - no p52 - TRAF3 dynamics like in the wild-type case upon stimulation
NIK$^{-/-}$	- abolished p100 processing - defect in p52 production - TRAF3 level stays high after stimulation	unpublished personal communication Z. B. Yilmaz	$k_3 = 0$ nmol·h^{-1}	- high p100 level; no processing inducible - no p52 - TRAF3 level decreases after stimulation
p100$^{-/-}$	- no p100 - no wild-type p52 - no defect in TRAF3 degradation	personal communication Z. B. Yilmaz	$k_{11} = 0$ nmol·h^{-1}	- no p100 - no p52 - no effect on the TRAF3 dynamics
p100 with mutated phosphoryla-tion site	- no phosphorylated p100 - blocked inducible p100 processing - no occurrence of p52	(Xiao et al., 2001);	$k_{13} = 0$ nmol^{-2}·h^{-1}	- no phophorylated p100 - no p100 processing; p100 level stays high - no p52
IKKα$^{-/-}$	- no p52 production - high p100 level	(Dejardin et al., 2002; Senftleben et al., 2001)		- no p52 - high p100 level
IKKα$^{-/-}$	- abolished p100 processing - defect in p52 production - no inducible change of the p100 or p52 level - RelB level is lower	(Yilmaz et al., 2003)	IKKα = 0 nM	- high p100 level; no processing inducible - no p52 - RelB level is only slightly decreased
IKKα that cannot be activated	- nuclear RelB level stays on a low level - nuclear p52 level stays on a low level	(Bonizzi et al., 2004)	$k_9 = 0$ nmol^{-1}·h^{-1}	- no p52/RelB^{nuc} - no increase upon pathway activation

The considered mutations affect distinct proteins involved in the signalling pathway: IKKα, NIK, p100, and TRAF3. Almost all effects of the investigated mutations can be reproduced by the model. Especially with respect to the effects on cytoplasmic p100 and nuclear p52, the model is able to reproduce the consequences of all mutations. This is a further indicator that the model describes the signalling branch very well.

The deletion of NIK (NIK$^{-/-}$), IKKα (IKKα$^{-/-}$) or p100 (p100$^{-/-}$) leads to the prevention of p100 processing. No p52 is produced as either the p100 processing is not induced or p100 is not available for its processing. This has been shown in experiments and is also reproduced by the model. The knock out of TRAF3 (TRAF3$^{-/-}$) facilitates the cleavage of p100. Its level decreases while the total level of p52 increases. If TRAF3 is missing in the model, the level of unbound NIK, which can be phosphorylated, is increased. Therefore, a higher amount of IKKα is phosphorylated and the processing of p100 is activated. Mutations preventing interactions of pathway components are: the TRAF3 mutation that is unable to bind NIK, and the NIK$^{aly/aly}$ mutation. If NIK carries the homozygotic alymphoplasia (aly) mutation (Shinkura et al., 1999), it is not able to bind IKKα (Luftig et al., 2001). In cells carrying these TRAF3 or NIK$^{aly/aly}$ mutations, p100 processing is prevented as the signal transduction is interrupted. This is also the case if the phosphorylation sites of IKKα and p100 are mutated or if NIK is not able to phosphorylate IKKα. These mutations inhibit the signal transduction from the NIK-TRAF-regulation module to the IKKα module or from the IKKα module to the p100-processing module.

Two experimental finding are not reproduced by the mathematical model. These are i) the TRAF3 dynamics in NIK$^{-/-}$-cells, and ii) the RelB concentration in IKKα$^{-/-}$-cells. In experiments, the TRAF3 dynamics differ in NIK$^{-/-}$-cells and in NIK$^{aly/aly}$-cells. In NIK$^{aly/aly}$-cells, the temporal behaviour of TRAF3 is similar to the dynamics in wild-type cells. In contrast, in NIK$^{-/-}$-cells, the TRAF3 dynamics is affected. A stimulus does not induce a decrease of the TRAF3 concentration. This observation is not reproduced by the simulations. In the model, TRAF3 is degraded after pathway stimulation even in the absence of NIK. This deviation between experiment and simulation reveals that not all regulatory mechanisms within the TRAF3-NIK-regulation module are included in the model. An extension of the module is reasonable and further discussed in section 3.2.8. The second deviation between the experimental observations and simulation results is the level of RelB if IKKα is mutated. The experimental data shows a strong decline in the RelB concentration, whereas in the model, the total concentration of RelB is only slightly decreased. A possible reason of this deviation is that IKKα is also involved in canonical NF-κB signalling that regulates the expression of RelB (Bren et al., 2001; Li et al., 1999 ; Scheidereit, 2006). Hence, the deletion of IKKα has an additional effect on the RelB concentration by affecting its production controlled by the canonical NF-κB signalling branch. So far, the crosstalk with the canonical NF-κB branch is not included, and thus, this effect is not captured by the model.

3.2.6   Sensitivity analysis of the model of non-canonical NF-κB signalling

A sensitivity analysis is performed to investigate the impact of all reactions on i) NIK^{PP}, ii) total cytoplasmic p100, and iii) nuclear p52/RelB. NIK^{PP} is very important in the transduction of the signal from the receptor to the downstream components. The dynamics of the precursor p100 is often considered to verify whether the non-canonical NF-κB pathway is active or not. Hence, it is of particular interest, which reactions have a strong impact on the concentration of cytoplasmic p100. Nuclear p52/RelB is the transcriptionally active pathway readout motivating the calculation of the sensitivity coefficients of this complex. The sensitivity coefficients are calculated as introduced in section 2.3.4. The rate constants are changed by 1%. For the three species, the effects on the unstimulated steady state and the stimulated steady state are examined. The calculated sensitivity coefficients are plotted as bar plots. The higher the value of the coefficient, the higher the impact of the particular reaction is. A positive value denotes that an increase of the parameter causes a higher steady state.

The sensitivity analysis is performed using the parameter set provided in Table 2. The values of the calculated sensitivity coefficients depend on the values of the kinetic parameters. Hence, one has to keep in mind that the usage of another parameter set would result in slightly different sensitivity coefficients. Additional sensitivity analyse with four other parameter sets show that the exact values of the sensitivity coefficients differ but not their signs. In other words: the strength of the effect would change if another parameter set is considered but not the types of effect (comparison not shown).

The sensitivity coefficients of NIK^{PP} (Figure 3.7), total cytoplasmic p100 (Figure 3.8) and nuclear p52/RelB (Figure 3.9) are shown.

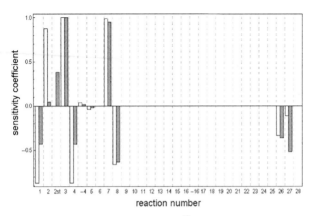

reaction number

Figure 3.7: Sensitivity coefficients of NIK^{PP}.

The sensitivity coefficients are calculated with respect to the unstimulated (white) and the stimulated steady state (grey). The numbers correspond to the reaction numbers shown in the model scheme in Figure 3.2. In the case of reversible reactions, the complex formation is denoted with a positive sign; the dissociation of the complex is labelled with the negative number.

Figure 3.7 reveals that only the reactions of the NIK-TRAF-regulation module have an impact on the steady states of NIK^{PP}. This is easy to understand since the NIK-TRAF-regulation module is not influenced by the downstream modules in the mathematical model. The type of effect of the reactions on the unstimulated and the stimulated steady state is the same, but differ partially in the strength. The production (reaction 3) and the phosphorylation of NIK (reaction 7) have a strong positive impact on both steady states. The constant degradation of TRAF3 (reaction 2) affects the unstimulated steady state strongly but has only a small influence on the stimulated steady state. On the contrary, the stimulus-dependent TRAF3 degradation (reaction 2st) only affects the stimulated steady state.

The steady state concentrations are negatively affected by the TRAF3 production (reaction 1), the formation of the TRAF3/NIK complex (reaction 4), the dephosphorylation of NIK (reaction 8), as well as the degradation of phosphorylated and unphosphorylated NIK (reactions 26 and 27, respectively). The production of TRAF3 (reaction 1) and the formation of the TRAF3/NIK complex (reaction 4) have the strongest effect on the unstimulated steady state. In contrast, the stimulated steady state is negatively affected the strongest by the dephosphorylation (reaction 8) and the degradation of NIK (reaction 27). The impact of the NIK dephosphorylation (reaction 8) and the degradation of NIK^{PP} (reaction 26) are similar on the unstimulated and stimulated steady state level of NIK^{PP}. The unstimulated steady state is negatively affected the strongest by the production of TRAF3 (reaction 1) and the formation of the TRAF3/NIK complex (reaction 4). The impact of both reactions is much smaller if the stimulated steady state is considered. The degradation of NIK (reaction 27) has a small impact on the unstimulated steady state but a high impact on the stimulated steady state. On the stimulated as well as unstimulated steady state, the impact of the dissociation of the TRAF3/NIK complex with and without NIK modification (reactions –4 and 5) is quite small.

In Figure 3.8 the sensitivity coefficients calculated for total cytoplasmic p100 are illustrated. Total cytoplasmic p100 takes into account cytoplasmic unmodified p100 and phosphorylated p100 (p100^{PP}). The sensitivity coefficients calculated for the single entities are shown in Appendix D, Figure D.1 and Figure D.2.

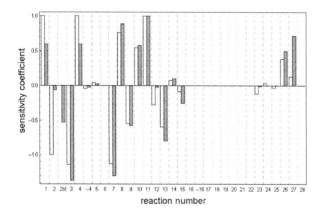

Figure 3.8: Sensitivity coefficients of total cytoplasmic p100.
Cytoplasmic p100 is the sum of unbound, unmodified cytoplasmic p100 and the phosphorylated species. The sensitivity coefficients are calculated with respect to the unstimulated (white) and the stimulated steady state levels (grey). The numbers correspond to the reaction numbers shown in the model scheme in Figure 3.2. In the case of reversible reactions, the complex formation is denoted with a positive sign; the dissociation of the complex is labelled with the negative number.

Figure 3.8 shows that the reactions of the NIK-TRAF-regulation module (reactions 1 to 8, 26 and 27) and the IKK module (reactions 9 and 10) affect the steady states of total cytoplasmic p100. Furthermore, cytoplasmic p100 is influenced by the reactions of the p100-processing module that are directly connected with p100. These are the production (reaction 11) and degradation of p100 (reaction 12), its phosphorylation and dephosphorylation (reactions 13 and 14, respectively) and the processing of phosphorylated p100 (reaction 15). Moreover, the reactions of the nucleo-cytoplasmic shuttle (reactions 23 and 24) as well as the $p100^{nuc}$ degradation (reaction 25) have an effect on the total concentration of cytoplasmic p100. The reactions including p52 and RelB (reactions 16 to 22 and 28) do not affect the steady states of total cytoplasmic p100.

The steady state concentration of total cytoplasmic p100 in the unstimulated as well as in the stimulated model is increased by a decreased supply of NIK^{PP}. The impact of the reactions of the NIK-TRAF-regulation module on p100 is of the opposite kind than the impact on the NIK^{PP} level (compare Figure 3.7 and Figure 3.8). The total cytoplasmic p100 concentration could also be increased by a reduced IKKα phosphorylation (reaction 9) or an enhanced $IKKα^{PP}$ dephosphorylation (reaction 10). A higher production rate of p100 (reaction 11) and a lower degradation rate (reaction 12) or phosphorylation rate (reaction 13) would contribute to a higher level of total cytoplasmic p100. Lowering the processing rate of p100 to p52 (reaction 15) increases the steady state of p100 as well. Regarding the impact on cytoplasmic p100, the shuttle into the nucleus has a negative influence on the steady state levels (reaction 23). This impact is quite small, just as the impact of the nuclear export of $p100^{nuc}$ (reaction 24) and the degradation of $p100^{nuc}$ (reaction 25).

Reactions of the TRAF-NIK-regulation module, the IKK module and in part the p100-processing module have a strong impact on the concentration of total cytoplasmic p100. The nuclear shuttle of p100 has only a minor effect and reactions taking p52 or RelB into account do not affect the steady states of cytoplasmic p100. In the model, the processing of $p100^{PP}$ and the subsequent production of p52 is included as an irreversible reaction. Hence, the reactions downstream of the p52 production do not affect p100.

Comparing the effects of total cytoplasmic p100, cytoplasmic unmodified p100 (Appendix D, Figure D.1) and phosphorylated p100 (Appendix D, Figure D.2) displays that the sensitivity coefficients of total cytoplasmic p100 are very similar to the sensitivity coefficients of unmodified cytoplasmic p100. Since the concentration of unmodified p100 is much higher than the concentration of phosphorylated p100, concentration changes in phosphorylated p100 have only little effects on the total cytoplasmic p100 concentration. Therefore, the sensitivity coefficients of total cytoplasmic p100 are mainly determined by the effects on unmodified cytoplasmic p100.

In Figure 3.9 the sensitivity coefficients of nuclear p52/RelB are shown.

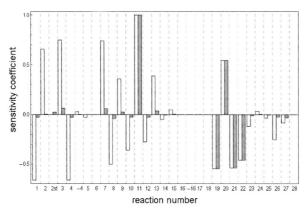

Figure 3.9: Sensitivity coefficients of nuclear p52/RelB.

The sensitivity coefficients are calculated with respect to the unstimulated (white) and the stimulated steady state (grey). The numbers correspond to the reaction numbers shown in the model scheme in Figure 3.2. In the case of reversible reactions, the complex formation is denoted with a positive sign; the dissociation of the complex is labelled with the negative number.

The sensitivity coefficients presented in Figure 3.9 illustrate that the unstimulated and the stimulated steady state of nuclear p52/RelB are affected by reactions of all modules. The unstimulated steady state concentration is strongly increased by increasing the constitutive degradation of TRAF3 (reaction 2), the production of NIK (reaction 3) or its phosphorylation (reaction 7). Furthermore, an enhanced phosphorylation of IKKα or p100 (reactions 9 or 13, respectively) results in a higher steady state of p52/RelB^{nuc}. This also results from an enhanced transport of p52/RelB into the nucleus (reaction 20). The strongest positive impact is mediated by the production of p100 (reaction 11). The unstimulated steady state of nuclear p52/RelB is strongly decreased by increasing the production of TRAF3 (reaction 1), the binding of TRAF3 and NIK (reaction 4), the dephosphorylation or the degradation of NIK^{PP} (reactions 8 and 26, respectively). The dephosphorylation of IKKα^{PP} (reaction 10) has a negative impact on the unstimulated steady state of nuclear p52/RelB. The degradation of p100 (reaction 12) has a negative influence as well as the degradation of p52/RelB in the cytoplasm (reaction 19) and in the nucleus (reaction 22), and the transport of nuclear p52/RelB out of the nucleus (reaction 21). The type of effect is equal for the stimulated steady state. The impact of the p100 production (reaction 11), the degradation of cytoplasmic p52/RelB (reaction 19), the nucleo-cytoplasmic shuttle of p52/RelB (reactions 20 and 21), and the degradation of p52/RelB^{nuc} (reaction 22) is equal for the unstimulated and the stimulated steady state. Except of these reactions and the stimulus-dependent TRAF3 degradation (reaction 2st), the impact of the reactions on the stimulated steady state is much lower than on the unstimulated steady state.

The reversible binding of p52 and RelB (reactions 16 and -16) affect the steady states of p52/RelB^{nuc} only marginally. The same holds for the degradation of p52 (reaction 28). This degradation rate is very small (Table 2). The binding of p52 and RelB prevents p52 to be degraded. Since the degradation rate is small, the impact of the binding to RelB is very small as well.

The steady state of nuclear p52/RelB depends on how much p52 is produced, hence, how much p100 is available to be processed. Additionally, the concentration of nuclear p52/RelB is determined by the degradation of cytoplasmic p52/RelB and the reactions of the p52/RelB-shuttle module.

Taken together, the sensitivity analyses emphasise the strong impact of the NIK-TRAF-regulation module on NIK^{PP} but also on the total cytoplasmic p100 and nuclear p52/RelB.

3.2.7   Discussion

A first mathematical model of the non-canonical NF-κB signalling branch has been developed (section 3.2.1). So far, modelling approaches concerning NF-κB have concentrated on the canonical signalling branch (Ashall et al., 2009; Basak et al., 2007; Cheong et al., 2008; Hoffmann et al., 2002; Kearns et al., 2006; Lipniacki et al., 2004; Mathes et al., 2008; Shih et al., 2009; Tay et al., 2010).

The developed model is based on the published experimental findings and the expertise of our collaboration partners Claus Scheidereit and Z. Buket Yilmaz.

The model is fitted to the experimental 12 h time course data that was measured in mouse embryonic fibroblasts (MEFs) treated with the non-canonical stimulus AC.H6 (LTβR stimulation) (section 3.2.2). A sensitivity analysis of the model is performed (section 3.2.6) and it is examined whether the mathematical model can reproduce the behaviour observed under numerous mutated conditions (section 3.2.5). It is planned to further validate the model. Simulated time courses caused by different stimuli (section 3.2.4) will be compared to experimental data.

Development of the model and parameter fit

The model includes the components TRAF3, NIK, IKKα, RelB, p100 and p52 and takes their molecular processes into account. The mathematical model reflects the experimental observations very well, including the decreasing concentration of TRAF3 and cytoplasmic p100 upon pathway activation and the stimulus-induced increase of the concentration of nuclear p52. Moreover, the model is able to reproduce the effects of numerous mutations (Table 3). The examined mutations affect IKKα, NIK, p100 and TRAF3 the major members of the non-canonical NF-κB signalling pathway. However, two details are not reflected by the model: On the one hand, the TRAF3 dynamic in NIK$^{-/-}$-cells, and on the other hand, the RelB level in IKKα$^{-/-}$-cells. The former deviation indicates that not all of the regulatory mechanisms within the NIK-TRAF-regulation module are included in the model (see section 3.2.8). A possible explanation for the difference in the RelB level is that RelB is a target gene of the canonical NF-κB branch (Bren et al., 2001). This branch is also affected by IKKα (Li et al., 1999 ; Scheidereit, 2006). Hence, to reflect all effects of an IKKα knockout on RelB, the effect on the canonical branch has to be considered as well. A model including both branches and their interactions on various levels will permit a more accurate prediction of the effects of an IKKα knockout.

The model is fitted to the experimental data by a Markov-Chain Monte Carlo method. As discussed in section 3.2.1, the fit is not unambiguous. One exemplary parameter set it selected for the analyses. Other reasonable parameter sets would result is other but similar simulated dynamics of the pathway components and sensitivity coefficients. However, the general conclusions would not change. Perturbation experiments and experiments focussing on the absolute concentrations of the pathway components are ongoing. This additional data will be integrated in the fitting procedure. This extended data set might be able to obtain an unambiguous parameter fit.

Predictions

The behaviour of the pathway components is predicted assuming a treatment with i) cycloheximide or ii) cycloheximide in combination with a non-canonical stimulus. The predicted time courses of TRAF3, NIK^{PP}, cytoplasmic p100 and nuclear p52 show that the proteins respond diversely to the inhibition of protein biosynthesis. The concentration of NIK decreases rapidly due to the rapid degradation of NIK (reaction −3). The protein is quite unstable. This was also shown in independent experiments in which NIK was only detectable upon administration of a proteasome inhibitor (Qing et al., 2005). In contrast to the fast decrease of NIK, the concentration of cytoplasmic p100 decreases to a lower degree. The precursor p100 is quite stable. This is in accordance with the experiments (data not shown) (Müller and Siebenlist, 2003). In general, the predicted dynamics of several pathway components upon application of cycloheximide and AC.H6 are in accordance with the experimental findings. However, further time course experiments are necessary for an extensive comparison between simulations and experimental data. The published experimental findings take only a few time points into account. Tighter time course data will better describe the dynamical behaviour. This will allow of the comparison of the simulated and measured dynamics in addition to the comparison of the steady state levels. The time course data will give additional information about the stability of the proteins and how this affects the signal transduction on different levels of the non-canonical NF-κB signalling pathway.

3.2.8   Outlook

The analyses of mutations of pathway components reveal that two effects of the mutation are not covered by the model. This leads to the conclusion that regulatory features within the NIK-TRAF-regulation module are lacking (section 3.2.5). NIK is a key regulator of the signalling pathway (Thu and Richmond, 2010). Recently, lots of experimental effort has been made to clarify the participating proteins and their interactions (Razani et al., 2010; Sanjo et al., 2010; Silke and Brink, 2010; Vallabhapurapu et al., 2008; Wallach and Kovalenko, 2008; Zarnegar et al., 2008). Based on these publications, a detailed qualitative model of the NIK-TRAF-regulation module is developed (Figure 3.10). It takes the published findings into account that focus on distinct aspects of the interactions and modifications of TRAF2, TRAF3, NIK, cIAP1/2 (referred to as cIAP in the context of complex notations), LTβR and IKKα.

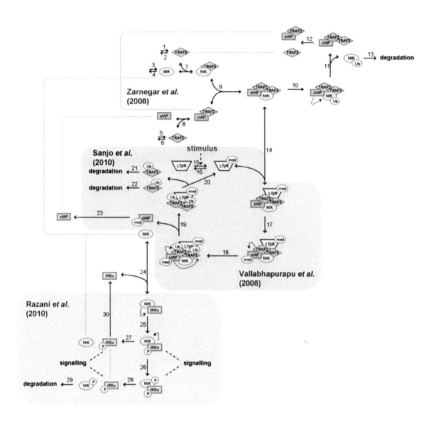

Figure 3.10: Detailed scheme of the regulation of the concentration of the major regulator NIK.

Based on the recent literature (Razani et al., 2010; Sanjo et al., 2010; Vallabhapurapu et al., 2008; Zarnegar et al., 2008), a reaction scheme concentrating on the regulation of the NIK concentration is proposed. Processes related to the publication by Zarnegar and colleagues are background-coloured in yellow, the green background highlights processes investigated by Vallabhapurapu and colleagues. The blue background marks reactions examined by Sanjo and co-workers while the purple background highlights the processes that are in the focus of Razani and colleagues. The scheme includes the components TRAF2, TRAF3, cIAP1/2 (referred to as cIAP), NIK, IKKα and the receptor LTβR.

The phosphorylation of a component is indicated by "P". Ubiquitinations and modifications are indicated by "Ub" and "mod", respectively. One-headed arrows denote reactions taking place in the indicated direction. Dashed arrows represent activation steps and double-headed arrows illustrate reversible binding reactions. Dotted intra-complex arrows indicate which molecule the modification of another protein mediates within the complex. Grey lines connect same entities. Dashed lines highlight IKKα^{PP}, the species that transduces the signal to downstream events. Numbers next to arrows denote the number of the particular reaction.

In this scheme, the mechanistic interactions suggested by Razani and colleagues (Razani et al., 2010), Sanjo and colleagues (Sanjo et al., 2010), Vallabhapurapu and colleagues (Vallabhapurapu et al., 2008) as well as Zarnegar and colleagues (Zarnegar et al., 2008) are combined: the formation of TRAF2/cIAP and TRAF3/NIK complexes that are able to bind, generating the four-component destruction complex (yellow background) (Zarnegar et al., 2008), the multimerisation of the receptor due to a stimulus and the subsequent dissociation of NIK and cIAP1/2 from the complex (blue background) (Sanjo et al., 2010). The proposed ubiquitination-cascade is included (green background) (Vallabhapurapu et al., 2008) and is assumed to take place if the complex is bound to the receptor. Thereby, the findings of Vallabhapurapu and colleagues and the findings of Sanjo and colleagues are combined. The regulation of NIK by IKKα discovered by Razani and colleagues (Razani et al., 2010) is included as well (purple background). The fate of the destruction complex under unstimulated conditions is unknown. In the model, it is assumed that after the release of ubiquitinated NIK, the TRAF2/TRAF3/cIAP complex dissociates into TRAF3 and the cIAP/TRAF2 complex. They may reenter into the system facilitating the degradation of further NIK molecules.

This scheme shows that the published findings can be combined to an overall picture of the mechanisms involved in the regulation of NIK. The single publications focus on distinct aspects but they are interconnected allowing the detailed description of the NIK-TRAF-regulation module. Based on the reaction scheme in Figure 3.10, a qualitative mathematical model is developed. It consists of 30 reactions and 25 components. TRAF3 is produced and degraded (reactions 1 and 2, respectively). The same holds for NIK (reactions 3 and 4) and TRAF2 (reactions 5 and 6). The pre-complexes TRAF3/NIK and TRAF2/cIAP are reversibly built by the single entities (reactions 7 and 8, respectively). The reversible binding of these two pre-complexes results in the formation of the destruction complex TRAF2/TRAF3/cIAP/NIK (reaction 9). Within this complex, NIK is ubiquitinated (reaction 10). Modified NIK is released from the complex (reaction 11) and degraded thereafter (reaction 13). The remaining complex TRAF2/TRAF3/cIAP dissociates into TRAF3 and TRAF2/cIAP (reaction 12). The complex TRAF2/TRAF3/cIAP/NIK reversibly binds to the modified LTβ receptor (reaction 14). The receptor is modified in presence of the stimulus (reaction 15) and can become unmodified again (reaction 16). The binding of the destruction complex and the receptor enables TRAF2 to modify cIAP (reaction 17). This leads to the ubiquitination of TRAF2 and TRAF3 (reaction 18). Subsequently, modified cIAP and NIK are released from the receptor-bound complex (reaction 19). The cIAP modification is a reversible process (reaction 23). Ubiquitinated TRAF2 and TRAF3 dissociate from the receptor (reaction 20) and are degraded (reactions 21 and 22, respectively). The kinase NIK reversibly binds IKKα (reaction 24) and activates IKKα by phosphorylation (reaction 25). Its phosphorylation enables $IKK\alpha^{PP}$ to phosphorylate NIK (reaction 26) leading to the degradation of phosphorylated NIK (reaction 29). This phosphorylation is different from the activating NIK phosphorylation considered in the model in section 3.2.1. The activating NIK phosphorylation is yet not included in the detailed mathematical model of the NIK-TRAF-regulation module. The complexes of NIK and IKKα dissociate into the single entities (reactions 27 and 28). $IKK\alpha^{PP}$ is the component that propagates the signal to the downstream events. $IKK\alpha^{PP}$ is dephosphorylated via reaction 30. The model equations are provided in Appendix D.3, equations D.46 to D.105.

The parameters are given in Table 13, Appendix D.3. Please, note that the parameters are not fitted. They are chosen in such a way, that the general temporal behaviour of TRAF3, TRAF2 and phosphorylated IKKα corresponds with the experimental observations. The respective concentration ranges are disregarded.

The initial analysis of the detailed model of the NIK-TRAF-regulation module reveals that the general temporal behaviour of TRAF2, TRAF3, NIK and phosphorylated IKKα can be reflected by the model. Corresponding to the experimental observations (Figure 3.3, unpublished data, (Thu and Richmond, 2010; Vallabhapurapu et al., 2008; Zarnegar et al., 2008)) the concentrations of TRAF2 and TRAF3 decrease upon pathway activation while the concentrations of NIK and $IKK\alpha^{PP}$ increase (plots not shown). This model also allows to investigate the effects of mutations of pathway components. The effects of mutations are summarised in Table 4. These mutations are also analysed in section 3.2.5. Here, only mutations of TRAF3 and NIK and the effects on these components are listed.

Table 4: Mutations of the NIK-TRAF-regulation module.

The mutations and their effects observed in experiments are listed with the respective references. Furthermore, the realisation in the model and the simulated effects are given. The considered mutations affect TRAF3 and NIK.

-/-: complete deletion of the respective gene, hence, a lack of the corresponding protein.

aly/aly: homozygotic alymphoplasia (aly) (Shinkura et al., 1999) mutation of the NIK gene. This mutation impedes the signal transduction from NIK to IKKα (Luftig et al., 2001).

Mutation	Experimental observation	Reference	Realisation in the model	Effects in the model
TRAF3[-/-]	- increased NIK level	(He et al., 2006) (Vallabhapurapu et al., 2008)	$k_1 = 0$ nmol·h^{-1}	- NIK level is increased
NIK[aly/aly]	- stimulus-induced TRAF3 degradation like in the wild-type case	unpublished personal communication Z. B. Yilmaz	$k_{24} = 0$ nmol^{-1}·h^{-1}	- TRAF3 dynamics similar to the wild-type dynamics after stimulation
NIK[-/-]	- TRAF3 level stays high after stimulation	unpublished personal communication Z. B. Yilmaz	$k_3 = 0$ nmol·h^{-1}	- TRAF3 level stays high upon stimulation

With the detailed model of the NIK-TRAF-regulation module one is able to reproduce the effects in TRAF$^{-/-}$-, NIK$^{-/-}$- and NIK$^{aly/aly}$-cells on the TRAF3 and NIK levels (Table 4). The detailed model reflects the effects that are also reflected by the model introduced in section 3.2.1. Furthermore, the detailed model of the NIK-TRAF-regulation module is able to cover the different consequences of a knockout of NIK and an aly mutation of NIK on the concentration of TRAF3. The model introduced in section 3.2.1 is not capable of reflecting these differences (see Table 3).

In the detailed model of the NIK-TRAF-regulation module, the degradation of TRAF3 depends on NIK. Only in the presence of NIK, the destruction complex TRAF2/TRAF3/cIAP/NIK is formed which binds to the activated receptor. Bound to the receptor, TRAF3 is modified facilitating its degradation. Hence, if NIK is not available, the stimulus is not able to induce a decrease of the TRAF3 concentration. The TRAF3 concentration remains on a high level. This corresponds to unpublished experimental data (personal communication Z. B. Yilmaz). It is assumed that the aly mutation does not affect the reactions of the formation of the destruction complex, its binding to the receptor and the subsequent modifications. The aly mutation prevents the binding of NIK and IKKα. Hence, upon pathway stimulation, the temporal behaviour of TRAF3 under NIK$^{aly/aly}$ conditions and wild-type conditions are very similar. Since mutated NIK cannot bind IKKα, more NIK is available to bind TRAF3. Thus, TRAF3 degradation is more effective under conditions of the aly mutation than under wild-type conditions. However, the signal transduction to downstream events is inhibited since IKKα is not activated because IKKα and NIK do not bind. This is in accordance with the experimental findings (personal communication Z. B. Yilmaz).

The initial analyses of the detailed model of the TRAF-NIK-regulation module reveal that the model is able to: i) generate the general dynamics of the components of this module as observed in the experiments, and ii) reproduce the effects of distinct mutations. Hence, the model reflects the experimental findings very well and will be further investigated.

Besides the analysis of the detailed model of the NIK-TRAF-regulation module, the crosstalk to the canonical NF-κB signalling branch is of high interest. Several mechanisms exist linking both branches. They include the usage of proteins by both branches, such as TRAF2 (Scheidereit, 2006), NIK (Thu and Richmond, 2010) and components of the IKK complex (Scheidereit, 2006). Furthermore, the gene expression of several pathway components is regulated by the signalling branches themselves (Bren et al., 2001; Lombardi et al., 1995; Sun et al., 1993). The induction of the non-canonical branch results in an activation of the canonical branch which precedes non-canonical activation. Moreover, the five NF-κB proteins may form complexes other than p50/RelA and p52/RelB (Ryseck et al., 1992), such as a p50/RelB complex. Hence, numerous crossregulatory mechanisms exist suggesting a complex network. Modelling both branches in combination will help to understand this complex network and the reasons of crossregulatory mechanism occurring at different levels of the pathways.

4 Conclusions and outlook

In signal transduction, various general mechanisms exist. One of them is the regulated proteolysis. There, key proteins of the signalling pathway are constantly degraded to prevent signal transduction in a resting cell. Two pathways that use the mechanism of regulated proteolysis are i) the canonical Wnt/β-catenin pathway and ii) the non-canonical NF-κB signalling branch. Although they have been objects of research for years, these pathways are still not completely understood. Therefore, theoretical approaches are used to contribute to gain deeper insights. In this thesis, mathematical models taking into account the molecular processes of the pathway components are developed and analysed with respect to their dynamical properties.

In chapter 2 of this work, the Wnt/β-catenin signalling pathway is investigated. Starting point of these theoretical analyses is the mathematical model developed by Lee and colleagues (Lee et al., 2003). This model is further modified and analysed focussing on the effects of distinct mutations on the pathway's properties (sections 2.2, 2.3 and 2.4), the effects of inhibitions (sections 2.2 and 2.4) and cell-type-specific properties (section 2.4).

The analyses of the effects of inhibitions deal with a number of different aspects. First it is shown that inhibitions affect the β-catenin steady state and β-catenin fold-change differently. The steady state is strongly affected by the destruction reactions. In contrast, the Dsh-dependent and Dsh-independent release of GSK3 from the complex APC/Axin/GSK3 have the strongest impact on the β-catenin fold-change. For small perturbations this has been shown by Goentoro and colleague who have introduced the concept that the β-catenin fold-change is the relevant pathway readout (Goentoro and Kirschner, 2009). Second and more importantly, the effects of single and double inhibitions on both the unstimulated β-catenin steady state and the fold-change are analysed in models with APC mutations of different strengths. The comparisons among each other and with the wild-type model show that with respect to the steady state the strength of the mediated effect depends on the strength of the APC mutation. In general it holds that the stronger the mutation is, the lower the impact of the degradation reactions and the higher the impact of the alternative β-catenin degradation. Regarding the fold-change, the strength and the type of effect can vary between the individual APC mutations. Third, the effects of inhibitions are also analysed in the cellular models which include the binding of β-catenin and E-cadherin as well as the Axin-2 feedback loop. The analysis reveals that the strength of the Axin-2 feedback loop strongly affects the strength of effects of the inhibitions. If the feedback loop is strong, the destruction reactions still have a high impact on the unstimulated β-catenin steady state even if APC is strongly mutated. In contrast, if the feedback loop is weak, those inhibitions have only weak effects like in the wild-type model. Taken together these analyses clearly emphasise that i) the readout of interest has to be specified – in particular the steady state or the fold-change – and ii) one has to be certain about the regulatory features of and the occurrence of mutated components in the particular model that is analysed. Does a weak or strong APC mutation exist? How strong is the feedback via Axin-2? These details determine whether an inhibition has a strong or a weak effect. Inhibitors that work efficiently under specific conditions are not necessarily effective under other conditions. The analyses of the inhibitions also emphasise the important role of the formation and dissociation of the APC/β-catenin complex. Especially if the APC concentration is very low or if the β-catenin concentration is very high, this complex formation strongly influences the β-catenin concentration. Under these conditions, the APC concentration is a limiting factor. As a consequence, only a very low amount of APC is available for the formation of the destruction complex. To compensate for the variations in the APC level, Lee and colleagues have supposed an APC-dependent regulation of the Axin degradation (Lee et al., 2003). The lower the APC concentration is, the less Axin is degraded. This increased Axin concentration promotes the formation of the APC/Axin complex and subsequently the formation of the destruction complex. Thus, it counteracts the effect of the APC/β-catenin complex. For prospective analyses it would be of interest to include this additional regulatory mechanism of the APC-regulated Axin degradation as well. To which extent could it compensate the effects of mutations?

The binding of APC and β-catenin plays also a central role when wild-type and mutated β-catenin coexist in one cell. Mutated β-catenin sequesters APC, resulting in a lower concentration of the destruction complex. The theoretical analyses show that an inhibitor affecting the binding of APC and β-catenin could be a potent drug in both the system with an APC mutation and the system with a β-catenin mutation. Until today it is experimentally impossible to manipulate this reaction exclusively. Recent studies have focussed on the interaction domains of APC to elucidate the contributions of the individual domains to the different binding processes. Based on these results it could be possible to find drugs that specifically affect the APC/β-catenin complex formation.

The analyses of the cellular models show that distinct temporal behaviours of β-catenin can be generated by the same general model structure if different concentrations of pathway components are assumed. There, the two newly included features – the feedback loop via Axin-2 and the binding of β-catenin and E-cadherin – have different effects that interact with each other. By the Axin-2 feedback, the concentration of β-catenin is decreased. Moreover, the feedback enables the decreasing phase of β-catenin that means that after the initial increase of the β-catenin concentration upon pathway activation the β-catenin concentration decreases until the stimulated steady state level is reached. E-cadherin buffers β-catenin leading to a prolonged increasing phase. Although the experimental observations reveal strong differences in the β-catenin dynamics of the two cell-types, it is possible to explain them by the same mathematical model. Hence, the different dynamics do not contradict the concept of conserved pathway structures in different cell-types or organisms. It is also shown that APC mutations have different effects in the two different cellular models. If the Axin-2 feedback loop is strong, the effects of APC mutations are weaker than in the model in which this feedback loop is weak. Regarding the β-catenin fold-change in the hepatocyte-like model even APC mutations of intermediate strength do not have an effect. This might be an indication for two important facts: i) It might be an explanation why particular mutations have been found especially in specific tumours while they have not been observed in tumours including other types of cells. While a specific mutation strongly affects the β-catenin dynamics in one cell-type, the conditions of another cell-type could compensate for the effects of that mutation. ii) It could be an indication that also in the mutated systems the β-catenin fold-change is the relevant pathway readout.

In summary, the theoretical analyses of the Wnt/β-catenin signalling pathway emphasise that it is important to be certain about the relevant pathway readout (steady state versus fold-change). They also underline the importance of the formation of the APC/β-catenin complex, especially under abnormal conditions. The theoretical investigations reveal that this complex formation is an important regulatory feature and could be an effective target process for drugs. Furthermore, the investigations show the importance of cell-type-specific information for making appropriate cell-type-specific predictions. So far, the modelling is based on the detailed kinetic measurements in *Xenopus* oocyte extracts (Lee et al., 2003). Detailed information about concentrations or kinetic parameters in other cell-types is still not available. However, new experimental methods like mass spectrometry are now applied to address this problem (Chen et al., 2010b). Of particular interest is measuring of the Axin concentration in other cell-types. Is it a common feature that this concentration is much lower than the concentrations of other pathway components or does this specifically occur in *Xenopus* oocyte extracts? If the latter is the case, the model simplifications that are based on this finding cannot be applied to the models of other cell-types without restrictions.

Chapter 3 focuses on the non-canonical NF-κB signalling pathway. For that pathway, a first mathematical model is developed that takes into account the molecular processes of core components of the pathway (section 3.2). The model represents experimental observations well. More importantly, it offers the ability to explain the experimental observations of several mutations, including mutations of TRAF3, NIK, IKKα and p100. This emphasises that the mathematical model reflects the details of the interactions of the pathway components very well. However, there are two deviations between experimental observations and the simulated behaviour. One deviation derives from the fact that the non-canonical signalling branch is considered exclusively and not in combination with the canonical NF-κB signalling branch. To obtain a detailed picture of the overall NF-κB signalling and to explain further mutations, a model including both the canonical and the non-canonical signalling branch has to be developed. The second deviation is caused by the simplified representation of the NIK-TRAF-regulation module. If recent experimental findings are taken into account in developing a detailed mathematical model of this module, the behaviour experimentally found in mutated cells can be reproduced by that model.

In summary, these results emphasise that although the kinetic parameters are not completely identified, the model can reproduce experimental findings and explain the behaviour when pathway components are mutated. Additional perturbation experiments will contribute to a refinement of the kinetic parameters. Furthermore, ongoing experimental work focuses on the measurement of concentrations of pathway components. This data will also be integrated into the model.

In conclusion, this work shows that mathematical modelling approaches contribute to a better understanding of the investigated signal transduction pathways.

Appendices

A   The wild-type model of Wnt/β-catenin signalling

A.1   Illustration of pathway modules

For the core-model developed by Lee and colleagues (Lee et al., 2003), the authors have calculated the control coefficients. Based on this analysis they have suggested a modularity of the model. The five modules that have been introduced by Lee and colleagues are illustrated in Figure A.1A. In Figure A.1B, the modules that have been introduced by Goentoro and colleague are presented (Goentoro and Kirschner, 2009). The basis of these three modules is a dimensional analysis. The Dsh module introduced by Lee and colleagues is equal to the input module introduced by Goentoro and colleague. All other modules differ.

A

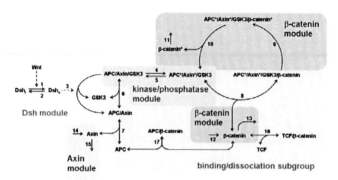

B

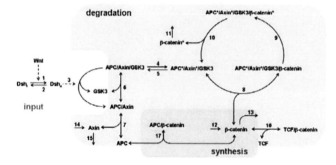

Figure A.1: Illustration of the modules in the core-model.

(A) The five modules that have been introduced by Lee and colleagues (Lee et al., 2003) are illustrated. These are the Dsh module (yellow background), the Axin module (brown background), the kinase/phosphatase module (dark green background), the β-catenin module (purple background) as well as the binding/dissociation subgroup (grey background). (B) The three modules are presented that have been published by Goentoro and colleague (Goentoro and Kirschner, 2009). In particular these are the input module (yellow background), the synthesis module (light green background) and the degradation module (blue background).

A.2   Model equations and parameters

Here, the equations of the wild-type model are given. Figure 2.4 shows a schematic representation of this model. In this scheme, the numbers next to arrows denote the number of the particular reaction. In the case of reversible reactions, the complex formation is denoted with a positive sign; the dissociation of the complex is labelled with the negative reaction number. Components in a complex are separated by a slash. The phosphorylation of a component is indicated by an asterisk.

Equations of the wild-type model of the Wnt/β-catenin pathway

$$\frac{d(Dsh_i)}{dt} = -v_1 + v_2 \tag{A.1}$$

$$\frac{d(Dsh_a)}{dt} = v_1 - v_2 \tag{A.2}$$

$$\frac{d(APC^*/Axin^*/GSK3)}{dt} = v_4 - v_5 - v_8 + v_{-8} + v_{10} \tag{A.3}$$

$$\frac{d(APC/Axin/GSK3)}{dt} = -v_3 - v_4 + v_5 + v_6 - v_{-6} \tag{A.4}$$

$$\frac{d(GSK3)}{dt} = v_3 - v_6 + v_{-6} \tag{A.5}$$

$$\frac{d(APC/Axin)}{dt} = v_3 - v_6 + v_{-6} + v_7 - v_{-7} \tag{A.6}$$

$$\frac{d(APC)}{dt} = -v_7 + v_{-7} - v_{17} + v_{-17} \tag{A.7}$$

$$\frac{d(APC^*/Axin^*/GSK3/\beta\text{-}catenin)}{dt} = v_8 - v_{-8} - v_9 \tag{A.8}$$

$$\frac{d(APC^*/Axin^*/GSK3/\beta\text{-}catenin^*)}{dt} = v_9 - v_{10} \tag{A.9}$$

$$\frac{d(\beta\text{-}catenin^*)}{dt} = v_{10} - v_{11} \tag{A.10}$$

$$\frac{d(\beta\text{-}catenin)}{dt} = -v_8 + v_{-8} + v_{12} - v_{13} - v_{16} + v_{-16} - v_{17} + v_{-17} \tag{A.11}$$

$$\frac{d(Axin)}{dt} = -v_7 + v_{-7} + v_{14} - v_{15} \tag{A.12}$$

$$\frac{d(TCF)}{dt} = -v_{16} + v_{-16} \tag{A.13}$$

$$\frac{d(TCF/\beta\text{-}catenin)}{dt} = v_{16} - v_{-16} \tag{A.14}$$

$$\frac{d(APC/\beta\text{-}catenin)}{dt} = v_{17} - v_{-17} \tag{A.15}$$

Rate equations of the wild-type model of the Wnt/β-catenin pathway

$$v_1 = k_1 \cdot Dsh_i \tag{A.16}$$

$$v_2 = k_2 \cdot Dsh_a \tag{A.17}$$

$$v_3 = k_3 \cdot Dsh_a \cdot (APC/Axin/GSK3) \tag{A.18}$$

$$v_4 = k_4 \cdot (APC/Axin/GSK3) \tag{A.19}$$

$$v_5 = k_5 \cdot (APC*/Axin*/GSK3) \tag{A.20}$$

$$v_6 = k_6 \cdot (APC/Axin) \cdot GSK3 \tag{A.21}$$

$$v_{-6} = k_{-6} \cdot (APC/Axin/GSK3) \tag{A.22}$$

$$v_7 = k_7 \cdot APC \cdot Axin \tag{A.23}$$

$$v_{-7} = k_{-7} \cdot (APC/Axin) \tag{A.24}$$

$$v_8 = k_8 \cdot (APC*/Axin*/GSK3) \cdot \beta\text{-}catenin \tag{A.25}$$

$$v_{-8} = k_{-8} \cdot (APC*/Axin*/GSK3/\beta\text{-}catenin) \tag{A.26}$$

$$v_9 = k_9 \cdot (APC*/Axin*/GSK3/\beta\text{-}catenin) \tag{A.27}$$

$$v_{10} = k_{10} \cdot (APC*/Axin*/GSK3/\beta\text{-}catenin*) \tag{A.28}$$

$$v_{11} = k_{11} \cdot \beta\text{-}catenin* \tag{A.29}$$

$$v_{12} = v_{12} \tag{A.30}$$

$$v_{13} = k_{13} \cdot \beta\text{-}catenin \tag{A.31}$$

$$v_{14} = v_{14} \tag{A.32}$$

$$v_{15} = k_{15} \cdot Axin \tag{A.33}$$

$$v_{16} = k_{16} \cdot TCF \cdot \beta\text{-}catenin \tag{A.34}$$

$$v_{-16} = k_{-16} \cdot (TCF/\beta\text{-}catenin) \tag{A.35}$$

$$v_{17} = k_{17} \cdot APC \cdot \beta\text{-}catenin \tag{A.36}$$

$$v_{-17} = k_{-17} \cdot (APC/\beta\text{-}catenin) \tag{A.37}$$

Kinetic parameters

In Table 5, the parameters are provided as they are used for the calculations in the wild-type model. They are based on the kinetic parameters published by (Lee et al., 2003). Table 6 contains the modified parameters if APC is mutated.

Table 5: Parameters of the wild-type model of the Wnt/β-catenin pathway.

Parameter	Value
k_1	1.82 min^{-1}
k_2	0.182 min^{-1}
k_3	0.5 nmol^{-1} min^{-1}
k_4	2.67 min^{-1}
k_5	1.33 min^{-1}
k_6	0.909 nmol^{-1} min^{-1}
k_{-6}	9.09 min^{-1}
k_7	10 nmol^{-1} min^{-1}
k_{-7}	500 min^{-1}
k_8	10 nmol^{-1} min^{-1}
k_{-8}	1200 min^{-1}
k_9	2060 min^{-1}
k_{10}	2060 min^{-1}
k_{11}	4.17 min^{-1}
v_{12}	4.23 nmol min^{-1}
k_{13}	0.00257 min^{-1}
v_{14}	1.67 nmol min^{-1}
k_{15}	0.000822 min^{-1}
k_{16}	10 nmol^{-1} min^{-1}
k_{-16}	300 min^{-1}
k_{17}	10 nmol^{-1} min^{-1}
k_{-17}	12000 min^{-1}

Table 6: Dissociation rates of complexes including APC in the wild-type model and models with APC mutations.

Following Cho and colleagues (Cho et al., 2006), the dissociation rates of the APC-containing complexes are affected if APC is mutated. In particular these are the dissociation of the APC/Axin complex (reaction −7), the release of β-catenin from the complex APC*/Axin*/GSK3/β-catenin (reaction −8) and the dissociation of the APC/β-catenin complex (reaction −17). For the sake of clarity, the respective parameter values of the wild-type model are given as well.

Mutation	Parameter	Value
wild-type		
	k_{-7}	500 min^{-1}
	k_{-8}	1200 min^{-1}
	k_{-17}	12000 min^{-1}
m1		
	k_{-7}	1682 min^{-1}
	k_{-8}	1200 min^{-1}
	k_{-17}	12000 min^{-1}
m5		
	k_{-7}	5800 min^{-1}
	k_{-8}	4007 min^{-1}
	k_{-17}	12000 min^{-1}
m10		
	k_{-7}	20000 min^{-1}
	k_{-8}	20000 min^{-1}
	k_{-17}	12000 min^{-1}
m13		
	k_{-7}	20000 min^{-1}
	k_{-8}	20000 min^{-1}
	k_{-17}	20000 min^{-1}

The basis of the parameter sets used in this thesis is the parameter set introduced by Lee and colleagues (Lee et al., 2003) and the modified parameter set published by Cho and colleagues (Cho et al., 2006). As discussed in section 2.2.1, in comparison to the parameters published by Lee and colleagues (Lee et al., 2003) the parameters of the wild-type model (Table 5) are multiplied by a factor of ten. Based on these new parameter values, the varied parameter values for the APC mutations (Table 6) are newly calculated following the assumptions of Cho and colleagues (Cho et al., 2006). These assumptions are shortly introduced in section 2.2.1.

Protein concentrations of proteins obeying the conservation relations

The total concentrations of proteins that fulfil the conservation relations are provided in Table 7. The concentrations are used as published by Lee and colleagues (Lee et al., 2003).

Table 7: Total concentration of proteins for which the conservation relations are assumed.

Component	Value
APC	100 nM
GSK3	50 nM
TCF	15 nM
Dsh	100 nM

B Wnt/β-catenin model including mutated β-catenin

B.1 Model equations and parameters

The equations of the model including mutated β-catenin are given in this section. Figure 2.12 shows a schematic representation of this model. There, the numbers next to arrows denote the number of the particular reaction. In the case of reversible reactions, the complex formation is denoted with a positive sign; the dissociation of the complex is labelled with the negative reaction number. Components in a complex are separated by a slash. The phosphorylation of a component is indicated by an asterisk.

Equations of the Wnt/β-catenin model including mutated β-catenin

Equations A.1, A.2, A.4–A.6, A.8–A.31 hold as listed in Appendix A. Equations A.3, A.7 and A.13 are replaced by equations B.1, B.2 and B.3, respectively. The additional model equations B.4 to B.14 are given here as well.

$$\frac{d(APC*/Axin*/GSK3)}{dt} = v_4 - v_5 - v_8 + v_{-8} + v_{10} - v_{20} + v_{-20} \tag{B.1}$$

$$\frac{d(APC)}{dt} = -v_7 + v_{-7} - v_{17} + v_{-17} - v_{22} + v_{-22} \tag{B.2}$$

$$\frac{d(TCF)}{dt} = -v_{16} + v_{-16} - v_{21} + v_{-21} \tag{B.3}$$

$$\frac{d(\beta\text{-}catenin^{mut})}{dt} = v_{18} - v_{19} - v_{20} + v_{-20} - v_{21} + v_{-21} - v_{22} + v_{-22} \tag{B.4}$$

$$\frac{d(TCF/\beta\text{-}catenin^{mut})}{dt} = v_{21} - v_{-21} \tag{B.5}$$

$$\frac{d(APC/\beta\text{-}catenin^{mut})}{dt} = v_{22} - v_{-22} \tag{B.6}$$

Additional rate equations of the Wnt/β-catenin pathway including mutated β-catenin

For reactions 1 to 17 the rate equations A.16–A.37 as listed in Appendix A are considered.

$$v_{18} = v_{18} \tag{B.7}$$

$$v_{19} = k_{19} \cdot \beta\text{-}catenin^{mut} \tag{B.8}$$

$$v_{20} = k_{20} \cdot (APC^*/Axin^*/GSK3) \cdot \beta\text{-}catenin^{mut} \tag{B.9}$$

$$v_{-20} = k_{-20} \cdot (APC^*/Axin^*/GSK3/\beta\text{-}catenin^{mut}) \tag{B.10}$$

$$v_{21} = k_{21} \cdot TCF \cdot \beta\text{-}catenin^{mut} \tag{B.11}$$

$$v_{-21} = k_{-21} \cdot (TCF/\beta\text{-}catenin^{mut}) \tag{B.12}$$

$$v_{22} = k_{22} \cdot APC \cdot \beta\text{-}catenin^{mut} \tag{B.13}$$

$$v_{-22} = k_{-22} \cdot (APC/\beta\text{-}catenin^{mut}) \tag{B.14}$$

Parameters

The parameters concerning wild-type β-catenin are used as provided in Table 5, Appendix A. The parameters concerning mutated β-catenin are given in Table 8. It is assumed that the mutation of β-catenin affects only its ability to be phosphorylated. It does not affect its ability or efficiency to interact with the binding partners. Therefore, the parameter values of analogue reactions are equal independent of whether they are related to wild-type or mutated β-catenin.

Table 8: Additional parameters of the mutated model of Wnt/β-catenin signalling.

Parameter	Value
v_{18}	4.23 nmol min^{-1}
k_{19}	0.00257 min^{-1}
k_{20}	10 nmol^{-1} min^{-1}
k_{-20}	1200 min^{-1}
k_{21}	10 nmol^{-1} min^{-1}
k_{-21}	300 min^{-1}
k_{22}	10 nmol^{-1} min^{-1}
k_{-22}	12000 min^{-1}

The total protein concentrations of APC, Dsh, TCF and GSK are not changed. The values of the wild-type model are used as given in Table 7, Appendix A.

B.2   Model equations and parameters of the minimal model

Here, the equations of the minimal models are given which are schematically presented in Figure 2.21. There, the numbers next to arrows denote the number of the particular reaction. In the case of reversible reactions, the complex formation is denoted with a positive sign; the dissociation of the complex is labelled with the negative reaction number. Components in a complex are separated by a slash.

Equations of the minimal wild-type model of Wnt/β-catenin signalling

$$\frac{d(\beta\text{-}catenin)}{dt} = v_1 - v_2 - v_3 - v_4 + v_{-4} \tag{B.15}$$

$$\frac{d(APC)}{dt} = -v_4 + v_{-4} \tag{B.16}$$

$$\frac{d(APC/\beta\text{-}catenin)}{dt} = v_4 - v_{-4} \tag{B.17}$$

Equations of the minimal mutated model

The equations B.15 and B.17 describe the behaviour of β-catenin and APC/β-catenin also in the mutated minimal model. Equation B.16 is replaced by B.18. Additional equations taking mutated β-catenin into account are listed below.

$$\frac{d(APC)}{dt} = -v_4 + v_{-4} - v_7 + v_{-7} \tag{B.18}$$

$$\frac{d(\beta\text{-}catenin^{mut})}{dt} = v_5 - v_6 - v_7 + v_{-7} \tag{B.19}$$

$$\frac{d(APC/\beta\text{-}catenin^{mut})}{dt} = v_7 - v_{-7} \tag{B.20}$$

Rate equations

$$v_1 = v_1 \tag{B.21}$$

$$v_2 = k_2 \cdot \beta\text{-}catenin \tag{B.22}$$

$$v_3 = k_3 \cdot \beta\text{-}catenin \cdot APC \tag{B.23}$$

$$v_4 = k_4 \cdot \beta\text{-}catenin \cdot APC \tag{B.24}$$

$$v_{-4} = k_{-4} \cdot (APC/\beta\text{-}catenin) \tag{B.25}$$

$$v_5 = v_5 \tag{B.26}$$

$$v_6 = k_6 \cdot \beta\text{-}catenin^{mut} \tag{B.27}$$

$$v_7 = k_7 \cdot \beta\text{-}catenin^{mut} \cdot APC \tag{B.28}$$

$$v_{-7} = k_{-7} \cdot (APC/\beta\text{-}catenin^{mut}) \tag{B.29}$$

Parameters

Table 9: Parameters of the minimal models.

The values are not adjusted. They are given in time units (tu) and concentration units (cu).

Parameter	Value
v_1	1 cu tu^{-1}
k_2	0.1 tu^{-1}
k_3	1 cu^{-1} tu^{-1}
k_3^{stim}	0.1 cu^{-1} tu^{-1}
k_4	1 cu^{-1} tu^{-1}
k_{-4}	10 tu^{-1}
v_5	1 cu tu^{-1}
k_6	0.1 tu^{-1}
k_7	1 cu^{-1} tu^{-1}
k_{-7}	10 tu^{-1}

According to the detailed model, the total concentration of APC does not change in time. It is fixed to 10 cu.

B.3   Effects of mutated β-catenin on the fold-change of wild-type β-catenin

In this section the effects of the coexistence of mutated and wild-type β-catenin on the fold-change of wild-type β-catenin are analysed. In sections 2.1 and 2.2 it is mentioned that the fold-change of β-catenin has been suggested to be the relevant readout of the Wnt/β-catenin signalling pathway (Goentoro and Kirschner, 2009). Therefore, it is of interest how the fold-change is affected by mutated β-catenin. Is it similar to the effects on the steady state? The analyses in section 2.2 show that APC mutations and inhibitions affect the steady state and fold-change differently. Do similarities to the effects mediated by APC mutations arise? What would be effective interventions that diminish the consequences of mutated β-catenin on the fold-change of wild-type β-catenin?

The fold-change of wild-type β-catenin is defined as the ratio of its stimulated steady state and its unstimulated steady state. Goentoro and Kirschner emphasise that the stimulated steady state results from a Wnt stimulus that is applied (Goentoro and Kirschner, 2009). Hence, in this section the pathway is activated by a Wnt stimulus. To be comparable with experiments in section 2.3, the pathway activation is simulated by the inhibition of the GSK3. Hence, the ratios of stimulated and unstimulated steady state presented in section 2.3 and fold-changes shown here differ.

Relation between the production of mutated β-catenin and the fold-change of wild-type β-catenin

The dependence of the wild-type β-catenin fold-change on the production of mutated β-catenin is investigated (Figure B.1). To this end, the fold-changes are calculated for increasing ratio f which is the ratio of the production rates of mutated and wild-type β-catenin (equation 2.5).

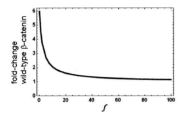

Figure B.1: Dependence of the fold-change of wild-type β-catenin on the ratio f.
The fold-change of wild-type β-catenin is plotted versus the ratio the ratio between the production rates of mutated and wild-type β-catenin (f). The production rate of wild-type β-catenin keeps its reference value.

Figure B.1 illustrates that a non-linear dependence of the fold-change of wild-type β-catenin on the production of mutated β-catenin exists. The more mutated β-catenin is produced, the lower is the inducible fold-change of wild-type β-catenin. As the availability of mutated protein influences the unstimulated steady state concentration of the wild-type protein (see Figure 2.13), the production rate of mutated protein affects the fold-change of wild-type β-catenin. The more mutated β-catenin is produced, the higher is the unstimulated steady state of wild-type β-catenin. The stimulated steady state is less strong affected. Consequently, a stimulus induced a smaller relative increase of the wild-type β-catenin concentration. The fold-change decreases. In general, this is similar to the findings of the effects of APC mutations on the β-catenin fold-change (see section 2.2.3).

Sensitivity coefficients of the wild-type β-catenin fold-change

To analyse the impact of the individual reactions on the wild-type β-catenin fold-change, the sensitivity coefficients of this characteristic are calculated (Figure B.2).

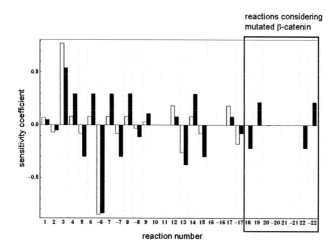

Figure B.2: Sensitivity coefficients of the fold-change of wild-type β-catenin.

The sensitivity coefficients of the fold-change of wild-type β-catenin are calculated in the wild-type model (white bars) and the mutated model (black bars). The parameters are changed by 1%. A positive coefficient implies that an increase of the respective parameter causes an increase of the wild-type β-catenin fold-change. The reactions taking mutated β-catenin into account are framed in blue. In the case of reversible reactions, the complex formation is denoted with a positive sign; the dissociation of the complex is labelled with the negative reaction number.

In the wild-type model (white bars in Figure B.2) the fold-change is mainly affected positively by the Dsh-dependent release of GSK3 from the complex APC/Axin/GSK3 (reaction 3). The main negative impact is mediated by the Dsh-independent release of GSK3 from this complex (reaction –6). The impact of all other parameters is much smaller. The production of β-catenin (reaction 12) has a positive influence on the fold-change while the alternative degradation (reaction 13) affects the fold-change negatively. Formation of the APC/β-catenin complex (process 17) has a positive effect on the fold-change. The dissociation of the complex (process –17) has a negative effect on that property (see also section 2.2 and (Goentoro and Kirschner, 2009)).

In the mutated model (black bars in Figure B.2), the impact of five of these six reactions decreases. The exception is the alternative degradation of wild-type β-catenin (reaction 13). Its impact increases. Except for the activation and deactivation of Dsh (reactions 1 and 2, respectively), the impact of all other reactions increases in the presence of mutated β-catenin as well. The reactions involving Dsh (reactions 1 to 3) represent the strength of the stimulus.

As the stimulus induces only a smaller fold-change in the mutated model (see Figure B.1), the impact of these reactions decreases. The impact of the production of wild-type β-catenin (reaction 12) decreases since the level of wild-type β-catenin is strongly increased due to the presence of mutated β-catenin. An additional slight increase of the production has only a little effect on the fold-change.

In addition to the reactions occurring in the wild-type model, reactions dealing with mutated β-catenin have an effect on the fold-change of the wild-type protein. The production of mutated β-catenin (reaction 18) as well as its binding to APC (reaction 22) have a negative effect on the fold-change of wild-type β-catenin. In contrast, the alternative degradation of the mutated protein (reaction 19) and the dissociation of the APC/β-catenin^{mut} complex affect the wild-type β-catenin fold-change positively. These reactions dealing with mutated β-catenin are the same reactions that affect the unstimulated steady state of wild-type β-catenin (Figure 2.18).

In contrast to the effects on the unstimulated β-catenin steady state, analogue reactions mediate the different type of effect on the fold-change (ratios of the sensitivity coefficients not shown). While for instance the activation of the formation of the APC/β-catenin complex causes a higher fold-change, the activation of the binding of APC and mutated β-catenin decreases the fold-change.

Based on the sensitivity analysis one can predict an appropriate perturbation of a destruction reaction (reactions 4 to 9, 14 and 15), with exception of reaction –6 (the Dsh-independent release of GSK3 from APC/Axin/GSK3), to be an effective intervention target to increase the wild-type β-catenin fold-change. These reactions have only little effects in the wild-type model but stronger effects in the mutated model. With respect to the unstimulated steady state, these reactions are not appropriate targets for interventions (Figure 2.18). Hence, like examined in the case of APC mutations, different reactions have to be intervened to diminish the effects on the β-catenin fold-change or the steady state level.

Minimal Model

The minimal model of the pathway (Figure 2.21) permits the analytical calculation of the fold-change of wild-type β-catenin. In the minimal model, the stimulus is simulated as a decrease of the rate constant k_3; $k_3 \rightarrow k_3^{stim}$, with $k_3^{stim} < k_3$. In the wild-type minimal model the following equation holds for the fold-change fc_w:

$$fc_w = \frac{(c_2 - c_3 - c_4^{stim} \cdot A) + \sqrt{(c_2 + c_3)^2 + (c_4^{stim} \cdot A)^2 + 2 \cdot c_4^{stim} \cdot A \cdot (-c_2 + c_3)}}{(c_2 - c_3 - c_4 \cdot A) + \sqrt{(c_2 + c_3)^2 + (c_4 \cdot A)^2 + 2 \cdot c_4 \cdot A \cdot (-c_2 + c_3)}}$$ (B.30)

with

$$c_1 = 2 \cdot k_2 k_4 ,$$

$$c_2 = k_1 k_4 ,$$

$$c_3 = k_2 k_{-4} ,$$

$$c_4 = k_3 k_{-4} , \text{ and}$$

$$c_4^{stim} = k_3^{stim} k_{-4}$$

with c_1, c_2, c_3, c_4, $c_4^{stim} > 0$, $c_4^{stim} < c_4$ and $A > 0$.

The fold-change of wild-type β-catenin in the mutated minimal model fc_m is given by:

$$fc_m = \frac{(c_2 - c_3 - c_4^{stim} \cdot A - c_5) + \sqrt{(c_2 + c_3)^2 + (c_4^{stim} \cdot A)^2 + 2 \cdot c_4^{stim} \cdot A \cdot (-c_2 + c_3) + c_5 \cdot (2 \cdot c_2 + 2 \cdot c_3 + 2 \cdot c_4^{stim} \cdot A + c_5)}}{(c_2 - c_3 - c_4 \cdot A - c_5) + \sqrt{(c_2 + c_3)^2 + (c_4 \cdot A)^2 + 2 \cdot c_4 \cdot A \cdot (-c_2 + c_3) + c_5 \cdot (2 \cdot c_2 + 2 \cdot c_3 + 2 \cdot c_4 \cdot A + c_5)}}$$

(B.31)

with $c_5 = \frac{k_2 k_{-4} k_5 k_7}{k_6 k_{-7}}$, $c_5 > 0$.

For all c_1, c_2, c_3, c_4, c_4^{stim}, $c_5 > 0$ and $A > 0$, it holds that $fc_w > 1$ and $fc_m > 1$, if $k_3^{stim} < k_3$. That means that the stimulated steady state of wild-type β-catenin is greater than its unstimulated steady state (see section 2.2).

B.4 Proofs

Positive β-catenin steady state in the minimal wild-type model

It is shown in general, that the steady state concentration of wild-type β-catenin in the minimal wild-type model (x_w) is greater than zero. Under the condition of positive parameter values it holds that A, c_1, c_2, c_3, $c_4 > 0$. x_w is calculated by:

$$x_w = \frac{1}{c_1}(c_2 - c_3 - c_4 \cdot A) + \frac{1}{c_1} \cdot \sqrt{(c_2 + c_3)^2 + (c_4 \cdot A)^2 + 2 \cdot c_4 \cdot A \cdot (-c_2 + c_3)} \; . \tag{B.32}$$

x_w always has a real solution as the following inequation holds:

$(c_2 + c_3)^2 + (c_4 \cdot A)^2 + 2 \cdot c_4 \cdot A \cdot (-c_2 + c_3) > 0$ (see section 2.3.5). Hence, the square root

$$\sqrt{(c_2 + c_3)^2 + (c_4 \cdot A)^2 + 2 \cdot c_4 \cdot A \cdot (-c_2 + c_3)} \tag{B.33}$$

is positive.

For $x_w > 0$, the inequation

$$-c_2 + c_3 + c_4 \cdot A < \sqrt{(c_2 + c_3)^2 + (c_4 \cdot A)^2 + 2 \cdot c_4 \cdot A \cdot (-c_2 + c_3)} \tag{B.34}$$

has to be fulfilled.

Here, one has to differentiate between two cases:

Case 1: $-c_2 + c_3 + c_4 \cdot A < 0$

As

$$0 < \sqrt{(c_2 + c_3)^2 + (c_4 \cdot A)^2 + 2 \cdot c_4 \cdot A \cdot (-c_2 + c_3)}$$

holds for all A, c_1, c_2, c_3, $c_4 > 0$, the inequation B.34 is fulfilled.

Case 2: $-c_2 + c_3 + c_4 \cdot A > 0$.

This allows to square the inequation B.34.

$$c_3^2 + (c_4 \cdot A)^2 + c_2^2 + 2 \cdot c_3 \cdot c_4 \cdot A - 2 \cdot c_2 \cdot c_3 - 2 \cdot c_2 \cdot c_4 \cdot A$$
$$< c_2^2 + c_3^2 + 2 \cdot c_2 \cdot c_3 + (c_4 \cdot A)^2 + 2 \cdot c_3 \cdot c_4 \cdot A - 2 \cdot c_2 \cdot c_4 \cdot A$$

Comparing the terms on both sides leads to:

$$-c_2 \cdot c_3 < c_2 \cdot c_3$$

This is always fulfilled for positive parameters.

Positive steady state of wild-type β-catenin in the minimal mutated model

One is also able to show that the steady state level of wild-type β-catenin in the mutated minimal model (x_m) is greater than zero for all A, c_1, c_2, c_3, c_4, $c_5 > 0$. x_m is calculated by:

$$x_m = \frac{1}{c_1}(c_2 - c_3 - c_4 \cdot A - c_5) + \frac{1}{c_1} \cdot \sqrt{(c_2 + c_3)^2 + (c_4 \cdot A)^2 + 2 \cdot c_4 \cdot A \cdot (-c_2 + c_3) + c_5 \cdot (2 \cdot c_2 + 2 \cdot c_3 + 2 \cdot c_4 \cdot A + c_5)} \tag{B.35}$$

x_m always has a real solution as

$$(c_2 + c_3)^2 + (c_4 \cdot A)^2 + 2 \cdot c_4 \cdot A \cdot (-c_2 + c_3) + c_5 \cdot (2 \cdot c_2 + 2 \cdot c_3 + 2 \cdot c_4 \cdot A + c_5) > 0 \quad \text{(see section 2.3.5).}$$

Thus, under the condition of positive parameters the square root

$$\sqrt{(c_2 + c_3)^2 + (c_4 \cdot A)^2 + 2 \cdot c_4 \cdot A \cdot (-c_2 + c_3) + c_5 \cdot (2 \cdot c_2 + 2 \cdot c_3 + 2 \cdot c_4 \cdot A + c_5)} \tag{B.36}$$

is always positive.

For $x_m > 0$ the inequation

$$-c_2 + c_3 + c_4 \cdot A + c_5 < \sqrt{(c_2 + c_3)^2 + (c_4 \cdot A)^2 + 2 \cdot c_4 \cdot A \cdot (-c_2 + c_3) + c_5 \cdot (2 \cdot c_2 + 2 \cdot c_3 + 2 \cdot c_4 \cdot A + c_5)} \tag{B.37}$$

has to be fulfilled.

Again, one has to differentiate between two cases.

Case 1: $-c_2 + c_3 + c_4 \cdot A + c_5 < 0$

It holds that

$$\sqrt{(c_2 + c_3)^2 + (c_4 \cdot A)^2 + 2 \cdot c_4 \cdot A \cdot (-c_2 + c_3) + c_5 \cdot (2 \cdot c_2 + 2 \cdot c_3 + 2 \cdot c_4 \cdot A + c_5)} > 0 .$$

Thus, the inequation B.37 is always fulfilled.

Case 2: $-c_2 + c_3 + c_4 \cdot A + c_5 > 0$.

This allows to square the inequation B.37.

$$c_2^2 + c_3^2 + (c_4 \cdot A)^2 + c_5^2 - 2 \cdot c_2 \cdot c_3 - 2 \cdot c_2 \cdot c_4 \cdot A - 2 \cdot c_2 \cdot c_5 + 2 \cdot c_3 \cdot c_4 \cdot A + 2 \cdot c_3 \cdot c_5 + 2 \cdot c_4 \cdot c_5 \cdot A$$
$$< c_2^2 + c_3^2 + 2 \cdot c_2 \cdot c_3 + (c_4 \cdot A)^2 + 2 \cdot c_3 \cdot c_4 \cdot A - 2 \cdot c_2 \cdot c_4 \cdot A + 2 \cdot c_2 \cdot c_5 + 2 \cdot c_3 \cdot c_5 + 2 \cdot c_4 \cdot c_5 \cdot A + c_5^2$$

Comparing the terms on both sides leads to

$$-(c_3 + c_5) < c_3 + c_5 .$$

This is fulfilled for all A, c_1, c_2, c_3, c_4, $c_5 > 0$.

Higher steady state level of wild-type β-catenin in the minimal mutated model than in the minimal wild-type model

In the following, the proof is shown that the steady state concentration of wild-type β-catenin in the minimal wild-type model (x_w) is smaller than in the minimal mutated model (x_m). One has to show that $x_w < x_m$ is fulfilled for all c_1, c_2, c_3, c_4, c_5, $A > 0$

$$x_w < x_m$$

$$\frac{1}{c_1}(c_2 - c_3 - c_4 \cdot A) + \frac{1}{c_1} \cdot \sqrt{(c_2 + c_3)^2 + (c_4 \cdot A)^2 + 2 \cdot c_4 \cdot A \cdot (-c_2 + c_3)}$$

$$< \frac{1}{c_1}(c_2 - c_3 - c_4 \cdot A - c_5) + \frac{1}{c_1} \cdot \sqrt{(c_2 + c_3)^2 + (c_4 \cdot A)^2 + 2 \cdot c_4 \cdot A \cdot (-c_2 + c_3) + c_5 \cdot (2 \cdot c_2 + 2 \cdot c_3 + 2 \cdot c_4 \cdot A + c_5)}$$

$$\sqrt{(c_2 + c_3)^2 + (c_4 \cdot A)^2 + 2 \cdot c_4 \cdot A \cdot (-c_2 + c_3)} + c_5$$

$$< \sqrt{(c_2 + c_3)^2 + (c_4 \cdot A)^2 + 2 \cdot c_4 \cdot A \cdot (-c_2 + c_3) + c_5 \cdot (2 \cdot c_2 + 2 \cdot c_3 + 2 \cdot c_4 \cdot A + c_5)}$$

Both sides of the inequation are positive. This allows to square the inequation.

$$2 \cdot c_5 \cdot \sqrt{(c_2 + c_3)^2 + (c_4 \cdot A)^2 + 2 \cdot c_4 \cdot A \cdot (-c_2 + c_3)} + c_5^2 + (c_2 + c_3)^2 + (c_4 \cdot A)^2 + 2 \cdot c_4 \cdot A \cdot (-c_2 + c_3)$$

$$< (c_2 + c_3)^2 + (c_4 \cdot A)^2 + 2 \cdot c_4 \cdot A \cdot (-c_2 + c_3) + c_5 \cdot (2 \cdot c_2 + 2 \cdot c_3 + 2 \cdot c_4 \cdot A + c_5)$$

$$\sqrt{(c_2 + c_3)^2 + (c_4 \cdot A)^2 + 2 \cdot c_4 \cdot A \cdot (-c_2 + c_3)} < (c_4 \cdot A + c_2 + c_3)$$

Again, both sides of the inequation are positive and it is squared.

$$(c_2 + c_3)^2 + (c_4 \cdot A)^2 + 2 \cdot c_4 \cdot A \cdot (-c_2 + c_3) < (c_2 + c_3)^2 + (c_4 \cdot A)^2 + 2 \cdot c_4 \cdot A \cdot (c_2 + c_3)$$

The comparison of the terms on both sides leads to:

$$-c_2 < c_2$$

This is always fulfilled for positive parameters.

Comparison of the stimulated and the unstimulated steady state levels in the minimal wild-type model

Furthermore, it can be shown in general that for all $c_1, c_2, c_3, c_4, A > 0$ the stimulated steady state concentration of wild-type β-catenin in the minimal wild-type model (x_w^{stim}) is greater than the unstimulated steady state (x_w).

For the stimulated steady state, $k_3 \rightarrow k_3^{stim}$ with $k_3^{stim} < k_3$. This leads to $c_4 \rightarrow c_4^{stim}$ with $c_4^{stim} < c_4$. This can also be described by $c_4^{stim} = c_4 - \alpha$ with $0 < \alpha < c_4$. This description is used in the following.

$$x_w < x_w^{stim}$$

$$\sqrt{(c_2 + c_3)^2 + (c_4 \cdot A)^2 + 2 \cdot c_4 \cdot A \cdot (-c_2 + c_3)}$$

$$< \alpha \cdot A + \sqrt{(c_2 + c_3)^2 + ((c_4 - \alpha) \cdot A)^2 + 2 \cdot (c_4 - \alpha) \cdot A \cdot (-c_2 + c_3)}$$

Since both sides of the inequation are positive it can be squared.

$$(c_2 + c_3)^2 + (c_4 \cdot A)^2 + 2 \cdot c_4 \cdot A \cdot (-c_2 + c_3)$$
$$< \alpha^2 \cdot A^2 + (c_2 + c_3)^2 + ((c_4 - \alpha) \cdot A)^2 + 2 \cdot (c_4 - \alpha) \cdot A \cdot (-c_2 + c_3)$$
$$+ 2 \cdot A \cdot \alpha \cdot \sqrt{(c_2 + c_3)^2 + ((c_4 - \alpha) \cdot A)^2 + 2 \cdot (c_4 - \alpha) \cdot A \cdot (-c_2 + c_3)}$$

This is simplified to:

$$A \cdot (c_4 - \alpha) + (-c_2 + c_3) \ < \ \sqrt{(c_2 + c_3)^2 + ((c_4 - \alpha) \cdot A)^2 + 2 \cdot (c_4 - \alpha) \cdot A \cdot (-c_2 + c_3)} \qquad \text{(B.38)}$$

In section 2.3.5 it was shown that the inequation

$$\sqrt{(c_2 + c_3)^2 + (c_4 \cdot A)^2 + 2 \cdot c_4 \cdot A \cdot (-c_2 + c_3)} > 0$$

holds, for all c_1, c_2, c_3, c_4, $A > 0$. This also holds for

$$\sqrt{(c_2 + c_3)^2 + ((c_4 - \alpha) \cdot A)^2 + 2 \cdot (c_4 - \alpha) \cdot A \cdot (-c_2 + c_3)} > 0, \qquad \text{(B.39)}$$

and $0 < \alpha < c_4$.

Hence, the square root exist and is always positive for c_1, c_2, c_3, c_4, $A > 0$ and $0 < \alpha < c_4$.

To investigate inequation B.38, the differentiation of two cases is needed.

Case 1: $A \cdot (c_4 - \alpha) + (-c_2 + c_3) < 0$ \qquad (B.40)

The inequation (B.38) is fulfilled as the square root

$$\sqrt{(c_2 + c_3)^2 + ((c_4 - \alpha) \cdot A)^2 + 2 \cdot (c_4 - \alpha) \cdot A \cdot (-c_2 + c_3)}$$

exists and is always positive.

Case 2: $A \cdot (c_4 - \alpha) + (-c_2 + c_3) > 0$ \qquad (B.41)

Both sides of inequation (B.38) are positive, allowing to square the inequation.

$$A \cdot (c_4 - \alpha) + (-c_2 + c_3) \ < \ \sqrt{(c_2 + c_3)^2 + ((c_4 - \alpha) \cdot A)^2 + 2 \cdot (c_4 - \alpha) \cdot A \cdot (-c_2 + c_3)}$$

$$(A \cdot (c_4 - \alpha))^2 + (-c_2 + c_3)^2 + 2 \cdot (A \cdot (c_4 - \alpha))(-c_2 + c_3) \ < \ (c_2 + c_3)^2 + ((c_4 - \alpha) \cdot A)^2 + 2 \cdot (c_4 - \alpha) \cdot A \cdot (-c_2 + c_3)$$

The comparison of the terms on both sides provides the following relation:

$$-2 \cdot c_2 \cdot c_3 \ < \ 2 \cdot c_2 \cdot c_3.$$

This is always fulfilled for positive parameters.

Comparison of the stimulated and the unstimulated steady state levels in the minimal mutated model

Also for the minimal mutated model one is able to show that the stimulated steady state level x_m^{stim} is always larger than the unstimulated steady state concentration x_m, $x_m < x_m^{stim}$.

$$\frac{1}{c_1}(c_2 - c_3 - c_4 \cdot A - c_5) + \frac{1}{c_1} \cdot \sqrt{(c_2 + c_3)^2 + (c_4 \cdot A)^2 + 2 \cdot c_4 \cdot A \cdot (-c_2 + c_3) + c_5 \cdot (2 \cdot c_2 + 2 \cdot c_3 + 2 \cdot c_4 \cdot A + c_5)}$$

$$< \frac{1}{c_1}(c_2 - c_3 - (c_4 - \alpha) \cdot A - c_5)$$

$$+ \frac{1}{c_1} \cdot \sqrt{(c_2 + c_3)^2 + ((c_4 - \alpha) \cdot A)^2 + 2 \cdot (c_4 - \alpha) \cdot A \cdot (-c_2 + c_3) + c_5 \cdot (2 \cdot c_2 + 2 \cdot c_3 + 2 \cdot (c_4 - \alpha) \cdot A + c_5)}$$

$$A \cdot (c_4 - \alpha) + (-c_2 + c_3) + c_5$$

$$< \sqrt{(c_2 + c_3)^2 + ((c_4 - \alpha) \cdot A)^2 + 2 \cdot (c_4 - \alpha) \cdot A \cdot (-c_2 + c_3) + c_5 \cdot (2 \cdot c_2 + 2 \cdot c_3 + 2 \cdot (c_4 - \alpha) \cdot A + c_5)}$$

$$\text{(B.42)}$$

In section 2.3.5 it was shown that

$$(c_2 + c_3)^2 + (c_4 \cdot A)^2 + 2 \cdot c_4 \cdot A \cdot (-c_2 + c_3) + c_5 \cdot (2 \cdot c_2 + 2 \cdot c_3 + 2 \cdot c_4 \cdot A + c_5) > 0$$

holds for all , c_2, c_3, c_4, c_5, $A > 0$. This also holds for $c_4 \rightarrow c_4^{stim}$ with $c_4^{stim} < c_4$, $c_4^{stim} = c_4 - \alpha$ with $0 < \alpha < c_4$. Hence,

$$\sqrt{(c_2 + c_3)^2 + ((c_4 - \alpha) \cdot A)^2 + 2 \cdot (c_4 - \alpha) \cdot A \cdot (-c_2 + c_3) + c_5 \cdot (2 \cdot c_2 + 2 \cdot c_3 + 2 \cdot (c_4 - \alpha) \cdot A + c_5)} > 0. \qquad \text{(B.43)}$$

Again one has to differentiate between two cases.

Case 1: $A \cdot (c_4 - \alpha) + (-c_2 + c_3) + c_5 < 0$

The inequation B.44 is fulfilled as the square root

$$\sqrt{(c_2 + c_3)^2 + ((c_4 - \alpha) \cdot A)^2 + 2 \cdot (c_4 - \alpha) \cdot A \cdot (-c_2 + c_3) + c_5 \cdot (2 \cdot c_2 + 2 \cdot c_3 + 2 \cdot (c_4 - \alpha) \cdot A + c_5)}$$

exists and is always positive.

Case 2: $A \cdot (c_4 - \alpha) + (-c_2 + c_3) + c_5 > 0$

Both sides of inequation B.44 are positive. Therefore, squaring the inequation is allowed.

$$(A \cdot (c_4 - \alpha) + (-c_2 + c_3) + c_5)^2$$

$$< (c_2 + c_3)^2 + ((c_4 - \alpha) \cdot A)^2 + 2 \cdot (c_4 - \alpha) \cdot A \cdot (-c_2 + c_3) + c_5 \cdot (2 \cdot c_2 + 2 \cdot c_3 + 2 \cdot (c_4 - \alpha) \cdot A + c_5)$$

Comparing the terms of both sides leads to the following relation:

$$-(c_3 + c_5) < c_3 + c_5$$

which is always fulfilled for positive parameters.

C   The cellular model of Wnt/β-catenin signalling

C.1   Model equations and parameters

Here, the equations of the cellular model are given which is schematically presented in Figure 2.29. There, the numbers next to arrows denote the number of the particular reaction. In the case of reversible reactions, the complex formation is denoted with a positive sign; the dissociation of the complex is labelled with the negative reaction number. Components in a complex are separated by a slash.

Equations of the cellular model of Wnt/β-catenin signalling

The equations A.1 to A.10 as well as A.13 to A15 are valid as listed in Appendix A. Equation A.11 and A.12 are replaced by the equations C.1 and C.2, respectively. Additional model equations are listed here.

$$\frac{d(\beta\text{-}catenin)}{dt} = -v_8 + v_{-8} + v_{12} - v_{13} - v_{16} + v_{-16} - v_{17} + v_{-17} - v_{23} + v_{-23} \tag{C.1}$$

$$\frac{d(Axin)}{dt} = -v_7 + v_{-7} + v_{14} - v_{15} + v_{19} \tag{C.2}$$

$$\frac{d(Axin\text{-}2\ mRNA)}{dt} = v_{18} - v_{20} \tag{C.3}$$

$$\frac{d(intermediate)}{dt} = v_{21} - v_{22} \tag{C.4}$$

$$\frac{d(E\text{-}cadherin)}{dt} = -v_{23} + v_{-23} \tag{C.5}$$

$$\frac{d(E\text{-}cadherin/\beta\text{-}catenin)}{dt} = v_{23} - v_{-23} \tag{C.6}$$

Additional rate equations of the cellular model of Wnt/β-catenin signalling

For reactions 1 to 17, the rate equations A.16 to A.37 are considered as listed in Appendix A. Additionally, equations C.7 to C.13 hold.

$$v_{18} = k_{18} \cdot (\beta\text{-}catenin + (TCF/\beta\text{-}catenin)) \tag{C.7}$$

$$v_{19} = k_{19} \cdot intermediate \tag{C.8}$$

$$v_{20} = k_{20} \cdot (Axin\text{-}2\ mRNA) \tag{C.9}$$

$$v_{21} = k_{21} \cdot (Axin\text{-}2\ mRNA) \tag{C.10}$$

$$v_{22} = k_{22} \cdot intermediate \tag{C.11}$$

$$v_{23} = k_{23} \cdot E\text{-}cadherin \cdot \beta\text{-}catenin \tag{C.12}$$

$$v_{-23} = k_{-23} \cdot (E\text{-}cadherin/\beta\text{-}catenin) \tag{C.13}$$

Kinetic parameters

The parameters used for both cell-type models are listed in Table 10. For the core-reactions (reactions 1 to 17), the kinetic parameters are given in Table 5, Appendix A.

Table 10: Additional kinetic parameters of the cellular model.

Parameter	Value
k_{18}	0.00005 nmol^{-1} min^{-1}
k_{19}	0.1 min^{-1}
k_{20}	0.1 min^{-1}
k_{21}	0.00075 min^{-1}
k_{23}	10 nmol^{-1} min^{-1}
k_{-23}	100 min^{-1}

The concentrations of GSK3, Dsh and TCF are not changed in comparison to the wild-type model. The values are listed in Table 7, Appendix A. The concentration of APC and the concentration of E-cadherin are considered to be different in the two different cell-types. Their total concentrations are given in Table 11 and Table 12 for the hepatocyte-like and C17.2-cell-like model, respectively. In addition, the parameter of the degradation rate of the intermediate is given in these Tables. It is assumed to be different in the two cellular models.

In the initial analyses of the cellular model, different total APC and E-cadherin levels as well as feedback strengths are used. They are provided in the corresponding descriptions of the analyses in section 2.4.

Table 11: Parameters and concentration of the hepatocyte-like model.

Parameter	Value
k_{22}	0.0005 min^{-1}
Concentration	
APC	100 nM
E-cadherin	2500 nM

Table 12: Parameters and concentration of the C17.2-cell-like model.

Parameter	Value
k_{22}	0.05 min^{-1}
Concentration	
APC	200 nM
E-cadherin	300 nM

C.2   Additional analyses of the cellular model

Dependence of the steady state concentration of total Axin on the APC level and the feedback strength

In Figure C.1 it is illustrated how the unstimulated steady state concentration of total Axin depends on the feedback strength and the total APC concentration. The stronger the feedback loop is, the higher the Axin concentration because more Axin is produced via the intermediate-dependent synthesis (reaction 19). In addition, the total concentration of Axin increases with increasing total concentrations of APC. The more APC is available, the more Axin is bound to APC. This facilitates the β-catenin degradation via the destruction cycle, and leads to a decrease in the activation of the feedback loop. Consequently, less Axin is produced in dependence of the intermediate. Nevertheless, the total Axin concentration is increased as a higher amount of Axin is bound in the complexes APC/Axin, APC/Axin/GSK, APC*/Axin*/GSK3, APC*/Axin*/GSK3/β-catenin and APC*/Axin*/GSK3/β-catenin*.

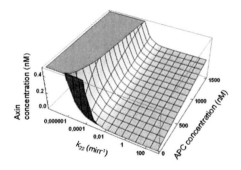

Figure C.1: Dependence of the total Axin concentration on the strength of the feedback and the total APC concentration.

The feedback strength is altered by the variation of the degradation rate of the intermediate (k_{22}). The k_{22}-values are plotted on a logarithmic scale. The total Axin concentration is cut off at 0.5 nM. In the extreme case of total $APC = 0$ nM and $k_{22} = 0.000001$ min^{-1}, Axin reaches the concentration of 3729 nM (not shown).

Effects of inhibitions on the β-catenin fold-change

In addition to the effects of inhibitions on the unstimulated steady state (Figure 2.47, section 2.4.7), the effects on the fold-change are analysed for the extract (Figure C.2A), the C17.2-cell-like (Figure C.2B) and the hepatocyte-like model (Figure C.2C). In row i, the wild-type case is shown, in row ii, row iii and row iv the APC mutations m1, m5 and m10, respectively, are presented. Figure C.2A is equal to row i to row iv in Figure 2.10, section 2.2.3.

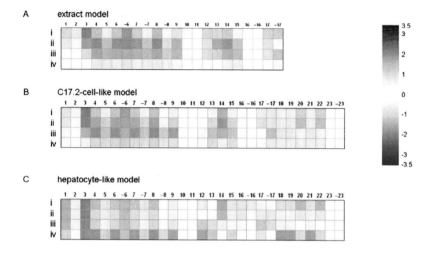

Figure C.2: Effects of inhibitions on the β-catenin fold-change in the extract and the cellular models.

The effects of an inhibition of reactions by 50% on the fold-change are examined in (A) the extract model, (B) the C17.2-cell-like model and (C) the hepatocyte-like model. The wild-type case (row i) and conditions of APC mutations m1 (row ii), m5 (row iii) as well as m10 (row iv) are considered. Parameters of the particular APC mutations are provided in Table 6, Appendix A.

White: no impact; green: an inhibition leads to a lower fold-change; red: an inhibition leads to a higher fold-change; the darker the colour, the higher the impact (see legend).

The numbers correspond to the reactions numbers shown in the model scheme in Figure 2.29. In the case of reversible reactions, the complex formation is denoted with a positive reaction number; the dissociation of the complex is labelled with the negative sign.

Comparing the wild-type cases in the extract and the cellular models reveals differences (comparing rows i in Figure C.2A, B and C). While in the extract model the inhibition of the β-catenin production (reaction 12) leads to a decrease of the fold-change, this inhibition causes almost no change of the fold-change in the C17.2-cell-like system and results in an increased fold-change in the hepatocyte-like system. As TCF participates in the feedback loop, perturbing the interaction of TCF and β-catenin (reactions 16 and –16) affects the fold-change if the feedback loop is present and works strongly like in the hepatocyte-like model

(Figure C.2C). In the C17.2-cell-like model, a perturbation of the interaction of β-catenin and TCF has only a small effect while it has no effect in the extract model.

In the extract model in the case of a weak APC mutation (m1, row ii), only one change of the type of effect occurs compared to the wild-type scenario (row i) which is related to the dissociation of the APC/β-catenin complex (reaction −17). In the C17.2-cell-like model and the hepatocyte-like model, the types of effect stay equal comparing the wild-type and the m1 models. However, the strength of the effect can change.

In the case of the intermediate APC mutation (m5, rows iii), more changes in the type of effect occur. Concentrating on the cellular models, these are especially the reactions of the feedback loop (reactions 18 to 22). Under wild-type conditions and in the presence of the weak APC mutation m1, the following holds for the reactions of the feedback loop: The inhibition of the degradation of the Axin-2 mRNA or the intermediate (reactions 20 and 22, respectively) leads to a higher concentration of the unstimulated β-catenin steady state (see Figure 2.47). The feedback loop is stronger activated. In contrast, the inhibition of the production of the Axin-2 mRNA (reaction 18), the intermediate-dependent Axin production (reaction 19), or the production of the intermediate (reaction 21), results in a lower unstimulated β-catenin steady state, and thus, in a feedback loop that is weaker activated. Independent of the specific inhibition, a stimulus induces an increase in the β-catenin concentration. However, if the initial β-catenin concentration is increased and the feedback loop works already strongly, the stimulus induces a lower relative concentration change than in the case of a lower β-catenin concentration and a weaker feedback loop. The fold-change decreases while starting from a lower β-catenin concentration and a weaker feedback loop results in an increased β-catenin fold-change. Due to stronger APC mutations (m5 and m10), less destruction complex is built. The unstimulated β-catenin steady state is strongly increased (see Figure 2.46, section 2.4.7), activating the feedback loop to a high degree. If for example the degradation of the intermediate is inhibited (reaction 22), this further increases the strength of the feedback loop. This increased feedback strength counteracts the lower concentration of available destruction complex that is caused by the strong APC mutation. More destruction complex is built that can be affected by the pathway stimulation. Hence, the system becomes more sensitive towards pathway stimulation. The fold-change increases. This line of argument also holds for the inhibition of the degradation rate of the Axin-2 mRNA (reaction 20). For the production of the Axin-2 mRNA (reaction 18), the production of the intermediate (reaction 21) and the intermediate-dependent Axin-2 production (reaction 19), the line of argument is the opposite way. Inhibiting these reactions cooperates with the effects of the APC mutation, resulting in a system that is more insensitive to its stimulation. The fold-change decreases. This explains the changes of the type of effect of the feedback loop reactions if one compares the effects in different models with APC mutations.

In the C17.2-cell-like model, the inhibition of the formation and dissociation of the APC/β-catenin complex (reaction 17 and −17) has different effects depending on the presence and strength of the APC mutation. This also occurs in the extract system (see section 2.2). Under wild-type and m1 conditions, an inhibition of the complex formation leads to a lower β-catenin fold-change. In contrast in the m5 and the m10 model, this inhibition leads to a higher fold-change. If the APC mutation is strong, the β-catenin concentration is high. This results in an increased formation of the APC/β-catenin complex, in which APC is sequestered. Thereby, less destruction complex is built and the stimulus is not able to induce a strong fold-change. If the formation of the APC/β-catenin complex is inhibited, this decrease of the APC concentration is decreased and a stimulus can induce a higher fold-change. This feature only occurs if the APC mutation is strong. A comparison with the hepatocyte-like system reveals, that this feature does not occur in this cellular model. In the hepatocyte-like model, the feedback loop is much stronger than in the C17.2-cell-like model. The strong feedback loop counteracts the effect of a strong APC mutation. More Axin is produced, facilitating the formation of the destruction complex, even in the presence of a strong APC mutation. Hence, a change of the type of effect of the formation and dissociation of the APC/β-catenin complex (reactions 17 and −17) occurs in the C17.2-cell-like but not in the hepatocyte-like model.

Last, the effects of an inhibition of the β-catenin production (reaction 12) are investigated. In the extract model, the inhibition of the β-catenin production decreases the fold-change in the wild-type and m1 model but increases the fold-change in the m5 and m10 model. This is discussed in section 2.2.3. In the C17.2-cell-like model, the impact of an inhibition of the β-catenin production (reaction 12) is quite low. It is different in the wild-type and m1 model in comparison to the m5 and m10 model. This phenomenon also occurs in the hepatocyte-like model, in which the impact of the β-catenin production is much higher than in the C17.2-cell-like model. However, in both cellular models, it holds that the inhibition of the β-catenin production (reaction 12) leads to an increased fold-change in the wild-type and m1 model but to a decreased fold-change in the m5 and m10 models. This is the opposite effect as observed in the extract model. While in the extract model under wild-type condition the inhibition of the β-catenin production leads to a lower fold-change as less β-catenin is produced (see also section 2.2.3), the inhibition of this reaction leads to an increase of the fold-change in the cellular models. In the cellular models, this inhibition causes two effects: i) it reduces the β-catenin concentration, but ii) subsequently it also reduces the strength of the feedback loop as less β-catenin is available for its induction. As seen in section 2.4.7, the stronger the feedback loop is, the lower the fold-change. By the inhibition of the β-catenin production, the feedback loop is weakened, enabling the induction of a higher fold-change. Due to the intermediate (m5) or the strong (m10) APC mutation, the β-catenin concentration is strongly increased reducing this former effect.

D Model of the non-canonical NF-κB signalling pathway

D.1 Model equations

Here, the equations of the model of the non-canonical NF-κB pathway are given which is schematically presented in Figure 3.2. There, the numbers next to arrows denote the number of the particular reaction. In the case of reversible reactions, the complex formation is denoted with a positive sign; the dissociation of the complex is labelled with the negative reaction number. Components in a complex are separated by a slash.

Equations of the model of non-canonical NF-κB signalling

$$\frac{d(TRAF3)}{dt} = v_1 - v_2 - v_4 + v_{-4} + v_5 \tag{D.1}$$

$$\frac{d(NIK)}{dt} = v_3 - v_{-3} - v_4 + v_{-4} - v_7 + v_8 - v_{27} \tag{D.2}$$

$$\frac{d(TRAF3/NIK)}{dt} = v_4 - v_{-4} - v_5 \tag{D.3}$$

$$\frac{d(NIK^{ub})}{dt} = v_5 - v_6 \tag{D.4}$$

$$\frac{d(NIK^{PP})}{dt} = v_7 - v_8 - v_{26} \tag{D.5}$$

$$\frac{d(IKK\alpha)}{dt} = -v_9 + v_{10} \tag{D.6}$$

$$\frac{d(IKK\alpha^{PP})}{dt} = v_9 - v_{10} \tag{D.7}$$

$$\frac{d(p100)}{dt} = v_{11} - v_{12} - v_{13} + v_{14} - v_{23} + v_{24} \tag{D.8}$$

$$\frac{d(p100^{PP})}{dt} = v_{13} - v_{14} - v_{15} \tag{D.9}$$

$$\frac{d(p52)}{dt} = v_{15} - v_{16} + v_{-16} - v_{28} \tag{D.10}$$

$$\frac{d(RelB)}{dt} = v_{17} - v_{18} - v_{16} + v_{-16} \tag{D.11}$$

$$\frac{\mathrm{d}(p52/RelB)}{\mathrm{d}t} = v_{16} - v_{-16} - v_{19} - v_{20} + v_{21} \tag{D.12}$$

$$\frac{\mathrm{d}(p100^{nuc})}{\mathrm{d}t} = v_{23} - v_{24} - v_{25} \tag{D.13}$$

$$\frac{\mathrm{d}(p52/RelB^{nuc})}{\mathrm{d}t} = v_{20} - v_{21} - v_{22} \tag{D.14}$$

Rate equations of the model of the non-canonical NF-κB pathway

$$v_1 = v_1 \tag{D.15}$$

$$v_2 = k_2 \cdot TRAF3 + k_2^{stim} \cdot TRAF3 \tag{D.16}$$

$$v_3 = v_3 \tag{D.17}$$

$$v_{-3} = k_{-3} \cdot NIK \tag{D.18}$$

$$v_4 = k_4 \cdot TRAF3 \cdot NIK \tag{D.19}$$

$$v_{-4} = k_{-4} \cdot TRAF3/NIK \tag{D.20}$$

$$v_5 = k_5 \cdot TRAF3/NIK \tag{D.21}$$

$$v_6 = k_6 \cdot NIK^{ub} \tag{D.22}$$

$$v_7 = k_7 \cdot NIK \tag{D.23}$$

$$v_8 = k_8 \cdot NIK^{PP} \tag{D.24}$$

$$v_9 = k_9 \cdot IKK\alpha \cdot NIK^{PP} \tag{D.25}$$

$$v_{10} = k_{10} \cdot IKK\alpha^{PP} \tag{D.26}$$

$$v_{11} = v_{11} \tag{D.27}$$

$$v_{12} = k_{12} \cdot p100 \tag{D.28}$$

$$v_{13} = k_{13} \cdot p100 \cdot IKK\alpha^{PP} \tag{D.29}$$

$$v_{14} = k_{14} \cdot p100^{PP} \tag{D.30}$$

$$v_{15} = k_{15} \cdot p100^{PP} \tag{D.31}$$

$$v_{16} = k_{16} \cdot p52 \cdot RelB \tag{D.32}$$

$$v_{-16} = k_{-16} \cdot p52/RelB \tag{D.33}$$

$$v_{17} = v_{17} \tag{D.34}$$

$$v_{18} = k_{18} \cdot RelB \tag{D.35}$$

$$v_{19} = k_{19} \cdot p52/RelB \tag{D.36}$$

$$v_{20} = k_{20} \cdot p52/RelB \qquad\qquad (D.37)$$

$$v_{21} = k_{21} \cdot p52/RelB^{nuc} \qquad\qquad (D.38)$$

$$v_{22} = k_{22} \cdot p52/RelB^{nuc} \qquad\qquad (D.39)$$

$$v_{23} = k_{23} \cdot p100 \qquad\qquad (D.40)$$

$$v_{24} = k_{24} \cdot p100^{nuc} \qquad\qquad (D.41)$$

$$v_{25} = k_{25} \cdot p100^{nuc} \qquad\qquad (D.42)$$

$$v_{26} = k_{26} \cdot NIK^{PP} \qquad\qquad (D.43)$$

$$v_{27} = k_{27} \cdot NIK \qquad\qquad (D.44)$$

$$v_{28} = k_{28} \cdot p52 \qquad\qquad (D.45)$$

D.2   Sensitivity analysis

Sensitivity analysis of unmodified and phosphorylated cytoplasmic p100

The comparison of the sensitivity analyses of total cytoplasmic p100 (section 3.2.6, Figure 3.8), unmodified cytoplasmic p100 (Figure D.1) and phosphorylated cytoplasmic p100 (Figure D.2) reveals that the sensitivity coefficients of total and unmodified cytoplasmic p100 are very similar while the coefficients of phosphorylated cytoplasmic p100 differ.

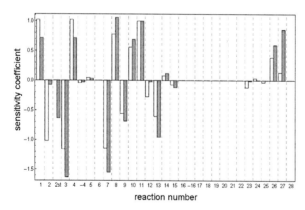

Figure D.1: Sensitivity coefficients of unmodified cytoplasmic p100.

The sensitivity coefficients are calculated with respect to the unstimulated (white) and the stimulated steady state (grey). The parameters are changed by 1%. The numbers correspond to the reaction numbers shown in the model scheme in Figure 3.2. In the case of reversible reactions, the complex formation is denoted with a positive sign; the dissociation of the complex is labelled with the negative number.

Figure D.1 shows that reactions of the i) NIK-TRAF-regulation, ii) IKK, and iii) p100-processing modules have a strong impact on the steady states of unmodified cytoplasmic p100. However, only the following reactions of the p100-processing module affect the steady state of unmodified cytoplasmic p100: i) its production (reaction 11), ii) the degradation of p100 (reaction 12), iii) the phosphorylation (reaction 13), iv) the dephosphorylation (reaction 14), as well as v) the processing of phosphorylated p100 (reaction 15). The reactions that include p52 do not affect the steady state of unmodified cytoplasmic p100 (reactions 16 to 19 and 28). Reactions of the p100 shuttle module have only a very small impact (reactions 23 to 25) while the reactions of the module of the p52 shuttle (reactions 20 to 22) do not affect the steady state at all. The calculated sensitivity coefficients have the same signs as the coefficients of total cytoplasmic p100. The exact values differ slightly since total cytoplasmic p100 takes also phosphorylated cytoplasmic p100 into account. The sensitivity coefficients of phosphorylated cytoplasmic p100, shown in Figure D.2, strongly differ from the sensitivity coefficients of unmodified cytoplasmic p100.

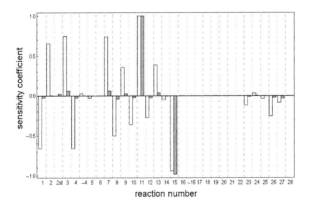

Figure D.2: Sensitivity coefficients phosphorylated cytoplasmic p100.
The sensitivity coefficients are calculated with respect to the unstimulated (white) and the stimulated steady state (grey). The parameters are changed by 1%. The numbers correspond to the reaction numbers shown in the model scheme in Figure 3.2. In the case of reversible reactions, the complex formation is denoted with a positive sign; the dissociation of the complex is labelled with the negative number.

Figure D.2 shows that the unstimulated steady state of phosphorylated cytoplasmic p100 (white bars) is affected by reactions of the i) NIK-TRAF-regulation module, ii) IKK module, iii) p100-processing module, and the iv) p100-shuttle module. Only the reactions of the p100-processing module that take into account p100 have an impact on the steady states (reactions 11 to 15). Reactions that are related to p52 or RelB (reactions 17 to 19 and 28) and reactions of the p52-shuttle module (reactions 20 to 22) have no impact on the steady states of phosphorylated cytoplasmic p100. Hence, the same reactions that affect the steady states of unmodified cytoplasmic p100 have an impact on the steady states of phosphorylated cytoplasmic p100. The comparison of Figure D.1 and Figure D.2 reveals that the reactions of the NIK-TRAF-regulation module and the reactions of the IKK module have the opposite type of effect on unmodified and on phosphorylated cytoplasmic p100. This means that a reaction that has a positive effect on the steady state of unmodified cytoplasmic p100 has a negative impact on the steady state of phosphorylated cytoplasmic p100. This is easy to understand since a signal coming from these upstream modules leads to the decrease of unmodified cytoplasmic p100 as it is phosphorylated, and at the same time to an increase of the concentration of phosphorylated cytoplasmic p100. Hence, the same reactions affect the steady state levels but in the opposite way.

D.3   Detailed model of the NIK-TRAF-regulation module

Here, the equations of the detailed model of the NIK-TRAF-regulation module are given which is schematically presented in Figure 3.10. There, the numbers next to arrows denote the number of the particular reaction. In the case of reversible reactions, the complex formation is denoted with a positive sign; the dissociation of the complex is labelled with the negative reaction number. Components in a complex are separated by a slash.

Equations of the detailed model of the NIK-TRAF-regulation module

$$\frac{d(TRAF3)}{dt} = v_1 - v_2 - v_7 + v_{-7} + v_{12} \tag{D.46}$$

$$\frac{d(NIK)}{dt} = v_3 - v_4 - v_7 + v_{-7} + v_{19} - v_{-19} - v_{24} + v_{-24} + v_{27} \tag{D.47}$$

$$\frac{d(cIAP)}{dt} = -v_8 + v_{-8} + v_{23} \tag{D.48}$$

$$\frac{d(TRAF2)}{dt} = v_5 - v_6 - v_8 + v_{-8} \tag{D.49}$$

$$\frac{d(TRAF3/NIK)}{dt} = v_7 - v_{-7} - v_9 + v_{-9} \tag{D.50}$$

$$\frac{d(TRAF2/cIAP)}{dt} = v_8 - v_{-8} - v_9 + v_{-9} + v_{12} \tag{D.51}$$

$$\frac{d(TRAF2/TRAF3/cIAP/NIK)}{dt} = v_9 - v_{-9} - v_{10} - v_{14} + v_{-14} \tag{D.52}$$

$$\frac{d(TRAF2/TRAF3/cIAP/NIK^{Ub})}{dt} = v_{10} - v_{11} \tag{D.53}$$

$$\frac{d(TRAF2/TRAF3/cIAP)}{dt} = v_{11} - v_{12} \tag{D.54}$$

$$\frac{d(NIK^{Ub})}{dt} = v_{11} - v_{13} \tag{D.55}$$

$$\frac{d(LT\beta R)}{dt} = -v_{15} + v_{16} \tag{D.56}$$

$$\frac{d(LT\beta R^{mod})}{dt} = -v_{14} + v_{-14} + v_{15} - v_{16} + v_{20} \tag{D.57}$$

$$\frac{d(LT\beta R^{mod}/TRAF2/TRAF3/cIAP/NIK)}{dt} = v_{14} - v_{-14} - v_{17} \tag{D.58}$$

$$\frac{d(LT\beta R^{mod}/TRAF2/TRAF3/cIAP^{mod}/NIK)}{dt} = v_{17} - v_{18} \tag{D.59}$$

$$\frac{d(LT\beta R^{mod}/TRAF2^{Ub}/TRAF3^{Ub}/cIAP^{mod}/NIK)}{dt} = v_{18} - v_{19} \tag{D.60}$$

$$\frac{d(LT\beta R^{mod}/TRAF2^{Ub}/TRAF3^{Ub})}{dt} = v_{19} - v_{20} \tag{D.61}$$

$$\frac{d(TRAF2^{Ub})}{dt} = v_{20} - v_{21} \tag{D.62}$$

$$\frac{d(TRAF3^{Ub})}{dt} = v_{20} - v_{22} \tag{D.63}$$

$$\frac{d(cIAP^{mod})}{dt} = v_{19} - v_{23} \tag{D.64}$$

$$\frac{d(IKK\alpha)}{dt} = -v_{24} + v_{-24} + v_{30} \tag{D.65}$$

$$\frac{d(IKK\alpha/NIK)}{dt} = v_{24} - v_{-24} - v_{25} \tag{D.66}$$

$$\frac{d(IKK\alpha^{PP}/NIK)}{dt} = v_{25} - v_{26} - v_{27} \tag{D.67}$$

$$\frac{d(IKK\alpha^{PP}/NIK^{PP})}{dt} = v_{26} - v_{28} \tag{D.68}$$

$$\frac{d(IKK\alpha^{PP})}{dt} = v_{27} + v_{28} - v_{30} \tag{D.69}$$

$$\frac{\mathrm{d}(NIK^{PP})}{\mathrm{d}t} = v_{28} - v_{29} \tag{D.70}$$

Rate equations

$$v_1 = v_1 \tag{D.71}$$

$$v_2 = k_2 \cdot TRAF3 \tag{D.72}$$

$$v_3 = v_3 \tag{D.73}$$

$$v_4 = k_4 \cdot NIK \tag{D.74}$$

$$v_5 = v_5 \tag{D.75}$$

$$v_6 = k_6 \cdot TRAF2 \tag{D.76}$$

$$v_7 = k_7 \cdot TRAF3 \cdot NIK \tag{D.77}$$

$$v_{-7} = k_{-7} \cdot (TRAF3/NIK) \tag{D.78}$$

$$v_8 = k_8 \cdot TRAF2 \cdot cIAP \tag{D.79}$$

$$v_{-8} = k_{-8} \cdot (TRAF2/cIAP) \tag{D.80}$$

$$v_9 = k_9 \cdot (TRAF3/NIK) \cdot (TRAF2/cIAP) \tag{D.81}$$

$$v_{-9} = k_{-9} \cdot (TRAF2/TRAF3/cIAP/NIK) \tag{D.82}$$

$$v_{10} = k_{10} \cdot (TRAF2/TRAF3/cIAP/NIK) \tag{D.83}$$

$$v_{11} = k_{11} \cdot (TRAF2/TRAF3/cIAP/NIK^{Ub}) \tag{D.84}$$

$$v_{12} = k_{12} \cdot (TRAF2/TRAF3/cIAP) \tag{D.85}$$

$$v_{13} = k_{13} \cdot NIK^{Ub} \tag{D.86}$$

$$v_{14} = k_{14} \cdot (TRAF2/TRAF3/cIAP/NIK) \tag{D.87}$$

$$v_{-14} = k_{-14} \cdot (LT\beta R^{mod}/TRAF2/TRAF3/cIAP/NIK) \tag{D.88}$$

$$v_{15} = k_{15} \cdot LT\beta R \tag{D.89}$$

$$v_{16} = k_{16} \cdot LT\beta R^{mod} \tag{D.90}$$

$$v_{17} = k_{17} \cdot (LT\beta R^{mod}/TRAF2/TRAF3/cIAP/NIK) \tag{D.91}$$

$$v_{18} = k_{18} \cdot (LT\beta R^{mod}/TRAF2/TRAF3/cIAP^{mod}/NIK) \tag{D.92}$$

$$v_{19} = k_{19} \cdot (LT\beta R^{mod}/TRAF2^{Ub}/TRAF3^{Ub}/cIAP^{mod}/NIK) \tag{D.93}$$

$$v_{20} = k_{20} \cdot (LT\beta R^{mod}/TRAF2^{Ub}/TRAF3^{Ub}) \tag{D.94}$$

$$v_{21} = k_{21} \cdot TRAF2^{Ub} \tag{D.95}$$

$$v_{22} = k_{22} \cdot TRAF3^{Ub} \tag{D.96}$$

$$v_{23} = k_{23} \cdot cIAP^{mod} \tag{D.97}$$

$$v_{24} = k_{24} \cdot NIK \cdot IKK\alpha \tag{D.98}$$

$$v_{-24} = k_{-24} \cdot (IKK\alpha/NIK) \tag{D.99}$$

$$v_{25} = k_{25} \cdot (IKK\alpha/NIK) \tag{D.100}$$

$$v_{26} = k_{26} \cdot (IKK\alpha^{PP}/NIK) \tag{D.101}$$

$$v_{27} = k_{27} \cdot (IKK\alpha^{PP}/NIK) \tag{D.102}$$

$$v_{28} = k_{28} \cdot (IKK\alpha^{PP}/NIK^{PP}) \tag{D.103}$$

$$v_{29} = k_{29} \cdot NIK^{PP} \tag{D.104}$$

$$v_{30} = k_{30} \cdot IKK\alpha^{PP} \tag{D.105}$$

Parameters

Table 13: Parameters of the detailed model of the NIK-TRAF-regulation module

Parameter	Value
v_1	10 nmol h^{-1}
k_2	1 h^{-1}
v_3	0.1 nmol h^{-1}
k_4	10 h^{-1}
v_5	10 nmol h^{-1}
k_6	1 h^{-1}
k_7	100 nmol^{-1} h^{-1}
k_{-7}	10 h^{-1}
k_8	100 nmol^{-1} h^{-1}
k_{-8}	10 h^{-1}
k_9	50 nmol^{-1} h^{-1}
k_{-9}	10 h^{-1}
k_{10}	50 h^{-1}
k_{11}	100 h^{-1}
k_{12}	1 h^{-1}

(Continued on page 218)

Continued Table 13.

Parameter	Value
k_{13}	10 h^{-1}
k_{14}	1000 $nmol^{-1}\,h^{-1}$
k_{-14}	10 h^{-1}
k_{15}	100 h^{-1}
k_{16}	1 h^{-1}
k_{17}	100 h^{-1}
k_{18}	100 h^{-1}
k_{19}	100 h^{-1}
k_{20}	100 h^{-1}
k_{21}	100 h^{-1}
k_{22}	100 h^{-1}
k_{23}	10 h^{-1}
k_{24}	1 $nmol^{-1}\,h^{-1}$
k_{-24}	0.1 h^{-1}
k_{25}	10 h^{-1}
k_{26}	10 h^{-1}
k_{27}	10 h^{-1}
k_{28}	10 h^{-1}
k_{29}	10 h^{-1}
k_{30}	10 h^{-1}

Table 14: Total concentrations of proteins obeying the conservation relations.

Component	Value
cIAP	10 nM
IKKα	20 nM
LTβR	50 nM

Abbreviations

APC	adenomatous polyposis coli
BAFF	B cell-activating factor
BLC	B lymphocyte chemokine
BMP	bone morphogenetic protein
β-TrCP	β-transducin-repeat-containing protein
CHX	cycloheximide
cIAP1/2	cellular inhibitor of apoptosis 1/2
CK1	casein kinase1
CYLD	cylindromatosis protein
Dkk	Dickkopf
Dsh	Dishevelled
EGF	epidermal growth factor
EMT	epithelial-mesenchymal transition
ERK	extracellular signal regulated-kinase
FGF	fibroblast growth factor
GAPDH	Glyceraldehyde 3-phosphate dehydrogenase
GSK3	glycogen synthase kinase-3
G protein	guanine nucleotide-binding protein
HCC	hepatocellular carcinoma
HMG	high mobility group
IKK	IκB kinase
IκB	inhibitor of κB binding
LEF	lymphocyte enhance factor
LRP5/6	low-density lipoprotein receptor-related protein 5/6
LTβR	lymphotoxin β receptor
Lys	Lysine
MCMC	Markov-Chain Monte Carlo
MAPK	mitogen-activated protein kinase, MAP kinase
MEF	mouse embryonic fibroblast
NEMO	NF-κB essential modifier
NF-κB	nuclear factor κ-light-chain-enhancer of activated B cells

NIK	NF-κB inducing kinase
ODE	ordinary differential equation
PCP	planar cell polarity
PCR	polymerase chain reaction
PDE	partial differential equation
PI3K	phosphoinositide 3-kinase
PP1	protein phosphatase 1
RANK	receptor activator of NF-κB
RIP	receptor interacting protein
RKIP	Raf kinase inhibitor protein
Ser	Serine
SUMO	small ubiquitin-like modifier
TCF	T-cell factor
Thr	Threonine
TRAF	TNF receptor-associated factor
TNF	Tumour necrosis factor

Bibliography

Aberle, H., Bauer, A., Stappert, J., Kispert, A. and Kemler, R. (1997). beta-catenin is a target for the ubiquitin-proteasome pathway. *Embo J* **16**, 3797-3804.

Albert, R. and Othmer, H. G. (2003). The topology of the regulatory interactions predicts the expression pattern of the segment polarity genes in Drosophila melanogaster. *J Theor Biol* **223**, 1-18.

Alon, U. (2007). Network motifs: theory and experimental approaches. *Nat Rev Genet* **8**, 450-461.

Amir, R. E., Haecker, H., Karin, M. and Ciechanover, A. (2004). Mechanism of processing of the NF-[kappa]B2 p100 precursor: identification of the specific polyubiquitin chain-anchoring lysine residue and analysis of the role of NEDD8-modification on the SCF[beta]-TrCP ubiquitin ligase. *Oncogene* **23**, 2540-2547.

Amit, S., Hatzubai, A., Birman, Y., Andersen, J. S., Ben-Shushan, E., Mann, M., Ben-Neriah, Y. and Alkalay, I. (2002). Axin-mediated CKI phosphorylation of beta-catenin at Ser 45: a molecular switch for the Wnt pathway. *Genes Dev* **16**, 1066-1076.

Angers, S. and Moon, R. T. (2009). Proximal events in Wnt signal transduction. *Nature Molecular Cell Biology* **10**, 468-477.

Ashall, L., Horton, C. A., Nelson, D. E., Paszek, P., Harper, C. V., Sillitoe, K., Ryan, S., Spiller, D. G., Unitt, J. F., Broomhead, D. S., Kell, D. B., Rand, D. A., See, V. and White, M. R. (2009). Pulsatile stimulation determines timing and specificity of NF-kappaB-dependent transcription. *Science* **324**, 242-246.

Aulehla, A. and Pourquié, O. (2008). Oscillating signaling pathways during embryonic development. *Current Opinion in Cell Biology* **20**, 632-637.

Aulehla, A., Wehrle, C., Brand-Saberi, B., Kemler, R., Gossler, A., Kanzler, B. and Herrmann, B. G. (2003). Wnt3a plays a major role in the segmentation clock controlling somitogenesis. *Dev Cell* **4**, 395-406.

Austinat, M., Dunsch, R., Wittekind, C., Tannapfel, A., Gebhardt, R. and Gaunitz, F. (2008). Correlation between beta-catenin mutations and expression of Wnt-signaling target genes in hepatocellular carcinoma. *Mol Cancer* **7**, 21.

Baldwin, A. S., Jr. (2001). Series introduction: the transcription factor NF-kappaB and human disease. *J Clin Invest* **107**, 3-6.

Barker, N. and Clevers, H. (2006). Mining the Wnt pathway for cancer therapeutics. *Nat Rev Drug Discov* **5**, 997-1014.

Basak, S. and Hoffmann, A. (2008). Crosstalk via the NF-kappaB signaling system. *Cytokine Growth Factor Rev* **19**, 187-197.

Basak, S., Kim, H., Kearns, J. D., Tergaonkar, V., O'Dea, E., Werner, S. L., Benedict, C. A., Ware, C. F., Ghosh, G., Verma, I. M. and Hoffmann, A. (2007). A fourth IkappaB protein within the NF-kappaB signaling module. *Cell* **128**, 369-381.

Basseres, D. S. and Baldwin, A. S. (2006). Nuclear factor-kappaB and inhibitor of kappaB kinase pathways in oncogenic initiation and progression. *Oncogene* **25**, 6817-6830.

Becker, V., Schilling, M., Bachmann, J., Baumann, U., Raue, A., Maiwald, T., Timmer, J. and KlingmÃ¼ller, U. (2010). Covering a Broad Dynamic Range: Information Processing at the Erythropoietin Receptor. *Science* **328**, 1404-1408.

Behar, M., Dohlman, H. G. and Elston, T. C. (2007a). Kinetic insulation as an effective mechanism for achieving pathway specificity in intracellular signaling networks. *Proc Natl Acad Sci U S A* **104,** 16146-16151.

Behar, M., Hao, N., Dohlman, H. G. and Elston, T. C. (2007b). Mathematical and Computational Analysis of Adaptation via Feedback Inhibition in Signal Transduction Pathways. *Biophys. J.* **93,** 806-821.

Behrens, J., Jerchow, B. A., Wurtele, M., Grimm, J., Asbrand, C., Wirtz, R., Kuhl, M., Wedlich, D. and Birchmeier, W. (1998). Functional interaction of an axin homolog, conductin, with beta-catenin, APC, and GSK3beta. *Science* **280,** 596-599.

Bienz, M. (2005). beta-Catenin: a pivot between cell adhesion and Wnt signalling. *Curr Biol* **15,** R64-67.

Bilic, J., Huang, Y.-L., Davidson, G., Zimmermann, T., Cruciat, C.-M., Bienz, M. and Niehrs, C. (2007). Wnt Induces LRP6 Signalosomes and Promotes Dishevelled-Dependent LRP6 Phosphorylation. *Science* **316,** 1619-1622.

Binder, B. and Heinrich, R. (2004). Interrelations between dynamical properties and structural characteristics of signal transduction networks. *Genome Informatics* **15,** 13-23.

Blüthgen, N., Legewie, S., Kielbasa, S. M., Schramme, A., Tchernitsa, O., Keil, J., Solf, A., Vingron, M., Schäfer, R., Herzel, H. and Sers, C. (2009). A systems biological approach suggests that transcriptional feedback regulation by dual-specificity phosphatase 6 shapes extracellular signal-related kinase activity in RAS-transformed fibroblasts. *FEBS Journal* **276,** 1024-1035.

Bonizzi, G., Bebien, M., Otero, D. C., Johnson-Vroom, K. E., Cao, Y., Vu, D., Jegga, A. G., Aronow, B. J., Ghosh, G., Rickert, R. C. and Karin, M. (2004). Activation of IKK[alpha] target genes depends on recognition of specific [kappa]B binding sites by RelB:p52 dimers. *EMBO Journal* **23,** 4202-4210.

Bonizzi, G. and Karin, M. (2004). The two NF-kappaB activation pathways and their role in innate and adaptive immunity. *Trends Immunol* **25,** 280-288.

Brabletz, S., Schmalhofer, O. and Brabletz, T. (2009). Gastrointestinal stem cells in development and cancer. *The Journal of Pathology* **217,** 307-317.

Braeuning, A., Ittrich, C., Kohle, C., Hailfinger, S., Bonin, M., Buchmann, A. and Schwarz, M. (2006). Differential gene expression in periportal and perivenous mouse hepatocytes. *Febs J* **273,** 5051-5061.

Brannon, M., Gomperts, M., Sumoy, L., Moon, R. T. and Kimelman, D. (1997). A beta-catenin/XTcf-3 complex binds to the siamois promoter to regulate dorsal axis specification in Xenopus. *Genes & Development* **11,** 2359-2370.

Bren, G. D., Solan, N. J., Miyoshi, H., Pennington, K. N., Pobst, L. J. and Paya, C. V. (2001). Transcription of the RelB gene is regulated by NF-kappaB. *Oncogene* **20,** 7722-7733.

Cadigan, K. M. and Liu, Y. I. (2006). Wnt signaling: complexity at the surface. *J Cell Sci* **119,** 395-402.

Cadigan, K. M. and Peifer, M. (2009). Wnt signaling from development to disease: insights from model systems. *Cold Spring Harbor Perspect Biol* **1,** a002881.

Carruba, G., Cervello, M., Miceli, M. D., Farruggio, R., Notarbartolo, M., Virruso, L., Giannitrapani, L., Gambino, R., Montalto, G. and Castagnetta, L. (1999). Truncated form of beta-catenin and reduced expression of wild-type catenins feature HepG2 human liver cancer cells. *Ann N Y Acad Sci* **886,** 212-216.

Chaves, M., Albert, R. and Sontag, E. D. (2005). Robustness and fragility of Boolean models for genetic regulatory networks. *J Theor Biol* **235,** 431-449.

Chen, B., Dodge, M. E., Tang, W., Lu, J., Ma, Z., Fan, C. W., Wei, S., Hao, W., Kilgore, J., Williams, N. S., Roth, M. G., Amatruda, J. F., Chen, C. and Lum, L. (2009a). Small molecule-mediated disruption of Wnt-dependent signaling in tissue regeneration and cancer. *Nat Chem Biol* **5**, 100-107.

Chen, M., Wang, J., Lu, J., Bond, M. C., Ren, X.-R., Lyerly, H. K., Barak, L. S. and Chen, W. (2009b). The Anti-Helminthic Niclosamide Inhibits Wnt/Frizzled1 Signaling. *Biochemistry* **48**, 10267-10274.

Chen, W., Chen, M. and Barak, L. S. (2010a). Development of small molecules targeting the Wnt pathway for the treatment of colon cancer: a high-throughput screening approach. *Am J Physiol Gastrointest Liver Physiol* **299**, G293-300.

Chen, Y., Gruidl, M., Remily-Wood, E., Liu, R. Z., Eschrich, S., Lloyd, M., Nasir, A., Bui, M. M., Huang, E., Shibata, D., Yeatman, T. and Koomen, J. M. (2010b). Quantification of beta-catenin signaling components in colon cancer cell lines, tissue sections, and microdissected tumor cells using reaction monitoring mass spectrometry. *J Proteome Res* **9**, 4215-4227.

Chen, Z., Venkatesan, A. M., Dehnhardt, C. M., Santos, O. D., Santos, E. D., Ayral-Kaloustian, S., Chen, L., Geng, Y., Arndt, K. T., Lucas, J., Chaudhary, I. and Mansour, T. S. (2009c). 2,4-Diamino-quinazolines as inhibitors of [beta]-catenin/Tcf-4 pathway: Potential treatment for colorectal cancer. *Bioorganic & Medicinal Chemistry Letters* **19**, 4980-4983.

Cheong, R., Hoffmann, A. and Levchenko, A. (2008). Understanding NF-kappaB signaling via mathematical modeling. *Mol Syst Biol* **4**, 192.

Cho, H. H., Song, J. S., Yu, J. M., Yu, S. S., Choi, S. J., Kim, D. H. and Jung, J. S. (2008). Differential effect of NF-[kappa]B activity on [beta]-catenin/Tcf pathway in various cancer cells. *FEBS Letters* **582**, 616-622.

Cho, K. H., Baek, S. and Sung, M. H. (2006). Wnt pathway mutations selected by optimal beta-catenin signaling for tumorigenesis. *FEBS Lett* **580**, 3665-3670.

Choi, J., Park, S. Y., Costantini, F., Jho, E. H. and Joo, C. K. (2004). Adenomatous polyposis coli is down-regulated by the ubiquitin-proteasome pathway in a process facilitated by Axin. *J Biol Chem* **279**, 49188-49198.

Cichowski, K. and Janne, P. A. (2010). Drug discovery: inhibitors that activate. *Nature* **464**, 358-359.

Claudio, E., Brown, K., Park, S., Wang, H. and Siebenlist, U. (2002). BAFF-induced NEMO-independent processing of NF-kappa B2 in maturing B cells. *Nat Immunol* **3**, 958-965.

Clevers, H. (2006). Wnt/beta-catenin signaling in development and disease. *Cell* **127**, 469-480.

Cong, F., Schweizer, L. and Varmus, H. (2004). Wnt signals across the plasma membrane to activate the {beta}-catenin pathway by forming oligomers containing its receptors, Frizzled and LRP. *Development* **131**, 5103-5115.

Coope, H. J., Atkinson, P. G., Huhse, B., Belich, M., Janzen, J., Holman, M. J., Klaus, G. G., Johnston, L. H. and Ley, S. C. (2002). CD40 regulates the processing of NF-kappaB2 p100 to p52. *Embo J* **21**, 5375-5385.

Courtois, G. and Gilmore, T. D. (2006). Mutations in the NF-kappaB signaling pathway: implications for human disease. *Oncogene* **25**, 6831-6843.

Daniels, D. L. and Weis, W. I. (2005). beta-catenin directly displaces Groucho/TLE repressors from Tcf/Lef in Wnt-mediated transcription activation. *Nat Struct Mol Biol* **12**, 364-371.

Daugherty, R. L. and Gottardi, C. J. (2007). Phospho-regulation of Beta-catenin adhesion and signaling functions. *Physiology (Bethesda)* **22**, 303-9.

Davidson, G., Wu, W., Shen, J., Bilic, J., Fenger, U., Stannek, P., Glinka, A. and Niehrs, C. (2005). Casein kinase 1 gamma couples Wnt receptor activation to cytoplasmic signal transduction. *Nature* **438**, 867-872.

de La Coste, A., Romagnolo, B., Billuart, P., Renard, C. A., Buendia, M. A., Soubrane, O., Fabre, M., Chelly, J., Beldjord, C., Kahn, A. and Perret, C. (1998). Somatic mutations of the beta-catenin gene are frequent in mouse and human hepatocellular carcinomas. *Proc Natl Acad Sci U S A* **95**, 8847-8851.

Dejardin, E. (2006). The alternative NF-kappaB pathway from biochemistry to biology: pitfalls and promises for future drug development. *Biochem Pharmacol* **72**, 1161-1179.

Dejardin, E., Droin, N. M., Delhase, M., Haas, E., Cao, Y., Makris, C., Li, Z.-W., Karin, M., Ware, C. F. and Green, D. R. (2002). The Lymphotoxin-[beta] Receptor Induces Different Patterns of Gene Expression via Two NF-[kappa]B Pathways. *Immunity* **17**, 525-535.

Dequeant, M. L., Glynn, E., Gaudenz, K., Wahl, M., Chen, J., Mushegian, A. and Pourquie, O. (2006). A complex oscillating network of signaling genes underlies the mouse segmentation clock. *Science* **314**, 1595-1598.

Dequeant, M.-L. and Pourquie, O. (2008). Segmental patterning of the vertebrate embryonic axis. **9**, 370-382.

Dimitrova, Y. N., Li, J., Lee, Y.-T., Rios-Esteves, J., Friedman, D. B., Choi, H.-J., Weis, W. I., Wang, C.-Y. and Chazin, W. J. (2010). Direct Ubiquitination of beta-Catenin by Siah-1 and Regulation by the Exchange Factor TBL1. *Journal of Biological Chemistry* **285**, 13507-13516.

Doble, B. W., Patel, S., Wood, G. A., Kockeritz, L. K. and Woodgett, J. R. (2007). Functional redundancy of GSK-3alpha and GSK-3beta in Wnt/beta-catenin signaling shown by using an allelic series of embryonic stem cell lines. *Dev Cell* **12**, 957-971.

Doble, B. W. and Woodgett, J. R. (2003). GSK-3: tricks of the trade for a multi-tasking kinase. *J Cell Sci* **116**, 1175-1186.

Dodge, M. E. and Lum, L. (2010). Drugging the Cancer Stem Cell Compartment: Lessons Learned from the Hedgehog and Wnt Signal Transduction Pathways. *Annu Rev Pharmacol Toxicol.*

Eklof Spink, K., Polakis, P. and Weis, W. I. (2000). Structural basis of the Axin-adenomatous polyposis coli interaction. *Embo J* **19**, 2270-2279.

Fearnhead, N. S., Britton, M. P. and Bodmer, W. F. (2001). The ABC of APC. *Hum Mol Genet* **10**, 721-733.

Fearon, E. R. (2009). PARsing the phrase "all in for Axin"- Wnt pathway targets in cancer. *Cancer Cell* **16**, 366-368.

Ferrell, J. E., Jr. (2002). Self-perpetuating states in signal transduction: positive feedback, double-negative feedback and bistability. *Curr Opin Cell Biol* **14**, 140-148.

Fitzgerald, J. B., Schoeberl, B., Nielsen, U. B. and Sorger, P. K. (2006). Systems biology and combination therapy in the quest for clinical efficacy. *Nat Chem Biol* **2**, 458-466.

Fritsche-Guenther, R., Witzel, F., Sieber, A., Herr, R., Schmidt, N., Braun, S., Brummer, T., Sers, C. and Blüthgen, N. (2011). Strong negative feedback from Erk to Raf confers robustness to MAPK signalling. *Mol Syst Biol* **7**, 489.

Fu, Z., Smith, P. C., Zhang, L., Rubin, M. A., Dunn, R. L., Yao, Z. and Keller, E. T. (2003). Effects of raf kinase inhibitor protein expression on suppression of prostate cancer metastasis. *J Natl Cancer Inst* **95**, 878-889.

Gaspar, C. and Fodde, R. (2004). APC dosage effects in tumorigenesis and stem cell differentiation. *Int J Dev Biol* **48**, 377-386.

Gebhardt, R., Baldysiak-Figiel, A., Krugel, V., Ueberham, E. and Gaunitz, F. (2007). Hepatocellular expression of glutamine synthetase: an indicator of morphogen actions as master regulators of zonation in adult liver. *Prog Histochem Cytochem* **41**, 201-266.

Gerondakis, S., Grumont, R., Gugasyan, R., Wong, L., Isomura, I., Ho, W. and Banerjee, A. (2006). Unravelling the complexities of the NF-kappaB signalling pathway using mouse knockout and transgenic models. *Oncogene* **25**, 6781-6799.

Geva-Zatorsky, N., Dekel, E., Cohen, A. A., Danon, T., Cohen, L. and Alon, U. (2010). Protein Dynamics in Drug Combinations: a Linear Superposition of Individual-Drug Responses. *Cell* **140**, 643-651.

Giles, R. H., van Es, J. H. and Clevers, H. (2003). Caught up in a Wnt storm: Wnt signaling in cancer. *Biochim Biophys Acta* **1653**, 1-24.

Goentoro, L. and Kirschner, M. W. (2009). Evidence that fold-change, and not absolute level, of beta-catenin dictates Wnt signaling. *Mol Cell* **36**, 872-884.

Goldbeter, A. and Pourquie, O. (2008). Modeling the segmentation clock as a network of coupled oscillations in the Notch, Wnt and FGF signaling pathways. *J Theor Biol* **252**, 574-585.

Götschel, F. (2008). Zelltypspezifische Analyse dynamischer Prozesse des Wnt/beta-Catenin Signalwegs mit Hilfe systembiologischer Verfahren. *Thesis*.

Götschel, F., Kern, C., Lang, S., Sparna, T., Markmann, C., Schwager, J., McNelly, S., von Weizsäcker, F., Laufer, S., Hecht, A. and Merfort, I. (2008). Inhibition of GSK3 differentially modulates NF-[kappa]B, CREB, AP-1 and [beta]-catenin signaling in hepatocytes, but fails to promote TNF-[alpha]-induced apoptosis. *Experimental Cell Research* **314**, 1351-1366.

Gwak, J., Song, T., Song, J.-Y., Yun, Y.-S., Choi, I.-W., Jeong, Y., Shin, J.-G. and Oh, S. (2009). Isoreserpine promotes [beta]-catenin degradation via Siah-1 up-regulation in HCT116 colon cancer cells. *Biochemical and Biophysical Research Communications* **387**, 444-449.

Haney, S., Bardwell, L. and Nie, Q. (2010). Ultrasensitive responses and specificity in cell signaling. *BMC Syst Biol* **4**, 119.

Hayden, M. S. and Ghosh, S. (2004). Signaling to NF-kappaB. *Genes Dev* **18**, 2195-2224.

Hayward, P., Brennan, K., Sanders, P., Balayo, T., DasGupta, R., Perrimon, N. and Martinez Arias, A. (2005). Notch modulates Wnt signalling by associating with Armadillo/beta-catenin and regulating its transcriptional activity. *Development* **132**, 1819-1830.

He, J. Q., Zarnegar, B., Oganesyan, G., Saha, S. K., Yamazaki, S., Doyle, S. E., Dempsey, P. W. and Cheng, G. (2006). Rescue of TRAF3-null mice by p100 NF-kappa B deficiency. *J Exp Med* **203**, 2413-2418.

He, X., Semenov, M., Tamai, K. and Zeng, X. (2004). LDL receptor-related proteins 5 and 6 in Wnt/beta-catenin signaling: arrows point the way. *Development* **131**, 1663-1677.

Hecht, A. and Kemler, R. (2000). Curbing the nuclear activities of beta-catenin. Control over Wnt target gene expression. *EMBO Rep* **1**, 24-28.

Hecht, A. and Stemmler, M. P. (2003). Identification of a promoter-specific transcriptional activation domain at the C terminus of the Wnt effector protein T-cell factor 4. *J Biol Chem* **278**, 3776-85.

Hecht, A., Vleminckx, K., Stemmler, M. P., van Roy, F. and Kemler, R. (2000). The p300/CBP acetyltransferases function as transcriptional coactivators of [beta]-catenin in vertebrates. *EMBO Journal* **19**, 1839-1850.

Heinrich, R., Neel, B. G. and Rapoport, T. A. (2002). Mathematical models of protein kinase signal transduction. *Mol Cell* **9**, 957-970.

Heinrich, R., Rapoport, S. M. and Rapoport, T. A. (1978). Metabolic regulation and mathematical models. *Progress in Biophysics and Molecular Biology* **32**, 1-82.

Henderson, B. R. and Fagotto, F. (2002). The ins and outs of APC and beta-catenin nuclear transport. *EMBO Rep* **3**, 834-839.

Heuberger, J. and Birchmeier, W. (2010). Interplay of Cadherin-Mediated Cell Adhesion and Canonical Wnt Signaling. *Cold Spring Harbor Perspectives in Biology* **2**, a002915.

Hoffmann, A., Levchenko, A., Scott, M. L. and Baltimore, D. (2002). The IkappaB-NF-kappaB signaling module: temporal control and selective gene activation. *Science* **298**, 1241-1245.

Hornberg, J. J., Bruggeman, F. J., Binder, B., Geest, C. R., de Vaate, A. J., Lankelma, J., Heinrich, R. and Westerhoff, H. V. (2005). Principles behind the multifarious control of signal transduction. ERK phosphorylation and kinase/phosphatase control. *Febs J* **272**, 244-258.

Hovanes, K., Li, T. W., Munguia, J. E., Truong, T., Milovanovic, T., Lawrence Marsh, J., Holcombe, R. F. and Waterman, M. L. (2001). Beta-catenin-sensitive isoforms of lymphoid enhancer factor-1 are selectively expressed in colon cancer. *Nat Genet* **28**, 53-57.

Huang, C. Y. and Ferrell, J. E., Jr. (1996). Ultrasensitivity in the mitogen-activated protein kinase cascade. *Proc Natl Acad Sci U S A* **93**, 10078-10083.

Huang, H. and He, X. (2008). Wnt/beta-catenin signaling: new (and old) players and new insights. *Curr Opin Cell Biol* **20**, 119-125.

Huang, S. M., Mishina, Y. M., Liu, S., Cheung, A., Stegmeier, F., Michaud, G. A., Charlat, O., Wiellette, E., Zhang, Y., Wiessner, S., Hild, M., Shi, X., Wilson, C. J., Mickanin, C., Myer, V., Fazal, A., Tomlinson, R., Serluca, F., Shao, W., Cheng, H., Shultz, M., Rau, C., Schirle, M., Schlegl, J., Ghidelli, S., Fawell, S., Lu, C., Curtis, D., Kirschner, M. W., Lengauer, C., Finan, P. M., Tallarico, J. A., Bouwmeester, T., Porter, J. A., Bauer, A. and Cong, F. (2009). Tankyrase inhibition stabilizes axin and antagonizes Wnt signalling. *Nature* **461**, 614-620.

Huber, O., Korn, R., McLaughlin, J., Ohsugi, M., Herrmann, B. G. and Kemler, R. (1996). Nuclear localization of beta-catenin by interaction with transcription factor LEF-1. *Mech Dev* **59**, 3-10.

Ikeda, S., Kishida, S., Yamamoto, H., Murai, H., Koyama, S. and Kikuchi, A. (1998). Axin, a negative regulator of the Wnt signaling pathway, forms a complex with GSK-3[beta] and [beta]-catenin and promotes GSK-3[beta]-dependent phosphorylation of [beta]-catenin. *EMBO Journal* **17**, 1371-1384.

Ille, F., Atanasoski, S., Falk, S., Ittner, L. M., Märki, D., Büchmann-Møller, S., Wurdak, H., Suter, U., Taketo, M. M. and Sommer, L. (2007). Wnt/BMP signal integration regulates the balance between proliferation and differentiation of neuroepithelial cells in the dorsal spinal cord. *Developmental Biology* **304**, 394-408.

Ingalls, P. B. and Sauro, H. M. (2003). Sensitivity analysis of stoichiometric networks: an extension of metabolic control analysis to non-steady state trajectories. *J Theor Biol* **222**, 23-36.

Jamora, C., DasGupta, R., Kocieniewski, P. and Fuchs, E. (2003). Links between signal transduction, transcription and adhesion in epithelial bud development. *Nature* **422**, 317-322.

Jensen, P. B., Pedersen, L., Krishna, S. and Jensen, M. H. (2010). A Wnt oscillator model for somitogenesis. *Biophys J* **98**, 943-950.

Jho, E.-h., Zhang, T., Domon, C., Joo, C.-K., Freund, J.-N. and Costantini, F. (2002). Wnt/{beta}-Catenin/Tcf Signaling Induces the Transcription of Axin2, a Negative Regulator of the Signaling Pathway. *Mol. Cell. Biol.* **22**, 1172-1183.

Ji, H., Wang, J., Nika, H., Hawke, D., Keezer, S., Ge, Q., Fang, B., Fang, X., Fang, D., Litchfield, D. W., Aldape, K. and Lu, Z. (2009). EGF-induced ERK activation promotes CK2-mediated disassociation of alpha-Catenin from beta-Catenin and transactivation of beta-Catenin. *Mol Cell* **36**, 547-559.

Kacser, H. and Burns, J. A. (1973). The control of flux. *Symp Soc Exp Biol* **27**, 65-104.

Karin, M. (2006). Nuclear factor-[kappa]B in cancer development and progression. *Nature* **441**, 431-436.

Karin, M. and Staudt, L. M. (2010). *NF-kappaB: A Network Hub Controlling Immunity, Inflammation, and Cancer.*

Kearns, J. D., Basak, S., Werner, S. L., Huang, C. S. and Hoffmann, A. (2006). IkappaBepsilon provides negative feedback to control NF-kappaB oscillations, signaling dynamics, and inflammatory gene expression. *J Cell Biol* **173**, 659-664.

Kestler, H. A. and Kuhl, M. (2008). From individual Wnt pathways towards a Wnt signalling network. *Philos Trans R Soc Lond B Biol Sci* **363**, 1333-1347.

Kholodenko, B. N. (2000). Negative feedback and ultrasensitivity can bring about oscillations in the mitogen-activated protein kinase cascades. *Eur J Biochem* **267**, 1583-1588.

Kiel, C. and Serrano, L. (2009). Cell type-specific importance of ras-c-raf complex association rate constants for MAPK signaling. *Sci Signal* **2**, ra38.

Kikuchi, A. (2003). Tumor formation by genetic mutations in the components of the Wnt signaling pathway. *Cancer Sci* **94**, 225-229.

Kim, D., Rath, O., Kolch, W. and Cho, K. H. (2007). A hidden oncogenic positive feedback loop caused by crosstalk between Wnt and ERK Pathways. *Oncogene*, 1-9.

Kimelman, D. and Xu, W. (2006). beta-catenin destruction complex: insights and questions from a structural perspective. *Oncogene* **25**, 7482-7491.

Klaus, A. and Birchmeier, W. (2008). Wnt signalling and its impact on development and cancer. *Nat Rev Cancer* **8**, 387-398.

Klipp, E. and Liebermeister, W. (2006). Mathematical modeling of intracellular signaling pathways. *BMC Neurosci* **7**, S10.

Klipp, E., Nordlander, B., Krüger, R., Gennemark, P. and Hohmann, S. (2005). Integrative model of the response of yeast to osmotic shock. *Nat Biotechnol* **23**, 975-982.

Kofahl, B. and Klipp, E. (2004). Modelling the dynamics of the yeast pheromone pathway. *Yeast* **21**, 831-850.

Kofahl, B. and Wolf, J. (2010). Mathematical modelling of Wnt/beta-catenin signalling. *Biochem Soc Trans* **38**, 1281-1285.

Kohler, E. M., Brauburger, K., Behrens, J. and Schneikert, J. (2010). Contribution of the 15 amino acid repeats of truncated APC to [beta]-catenin degradation and selection of APC mutations in colorectal tumours from FAP patients. **29**, 1663-1671.

Kohler, E. M., Chandra, S. H. V., Behrens, J. and Schneikert, J. (2009). {beta}-Catenin degradation mediated by the CID domain of APC provides a model for the selection of

APC mutations in colorectal, desmoid and duodenal tumours. *Hum. Mol. Genet.* **18,** 213-226.

Kohler, E. M., Derungs, A., Daum, G., Behrens, J. r. and Schneikert, J. (2008). Functional definition of the mutation cluster region of adenomatous polyposis coli in colorectal tumours. *Human Molecular Genetics* **17,** 1978-1987.

Komarova, N. L., Zou, X., Nie, Q. and Bardwell, L. (2005). A theoretical framework for specificity in cell signaling. *Mol Syst Biol* **1,** 2005 0023.

Krieghoff, E., Behrens, J. and Mayr, B. (2006). Nucleo-cytoplasmic distribution of beta-catenin is regulated by retention. *J Cell Sci* **119,** 1453-1463.

Krüger, R. and Heinrich, R. (2004). Model reduction and analysis of robustness for the Wnt/beta-catenin signal transduction pathway. *Genome Inform Ser Workshop Genome Inform* **15,** 138-48.

Kumar, A., Takada, Y., Boriek, A. M. and Aggarwal, B. B. (2004). Nuclear factor-kappaB: its role in health and disease. *J Mol Med (Berl)* **82,** 434-48.

Kuphal, F. and Behrens, J. (2006). E-cadherin modulates Wnt-dependent transcription in colorectal cancer cells but does not alter Wnt-independent gene expression in fibroblasts. *Experimental Cell Research* **312,** 457-467.

Laurent-Puig, P., Legoix, P., Bluteau, O., Belghiti, J., Franco, D., Binot, F., Monges, G., Thomas, G., Bioulac-Sage, P. and Zucman-Rossi, J. (2001). Genetic alterations associated with hepatocellular carcinomas define distinct pathways of hepatocarcinogenesis. *Gastroenterology* **120,** 1763-1773.

Lavergne, E., Hendaoui, I., Coulouarn, C., Ribault, C., Leseur, J., Eliat, P.-A., Mebarki, S., Corlu, A., Clement, B. and Musso, O. (2010). Blocking Wnt signaling by SFRP-like molecules inhibits in vivo cell proliferation and tumor growth in cells carrying active [beta]-catenin. *Oncogene* **30,** 423-433.

Lee, E., Salic, A., Krüger, R., Heinrich, R. and Kirschner, M. W. (2003). The roles of APC and Axin derived from experimental and theoretical analysis of the Wnt pathway. *PLoS Biol* **1,** 116-132.

Lehar, J., Stockwell, B. R., Giaever, G. and Nislow, C. (2008). Combination chemical genetics. *Nat Chem Biol* **4,** 674-681.

Lehar, J., Zimmermann, G. R., Krueger, A. S., Molnar, R. A., Ledell, J. T., Heilbut, A. M., Short, G. F., Giusti, L. C., Nolan, G. P., Magid, O. A., Lee, M. S., Borisy, A. A., Stockwell, B. R. and Keith, C. T. (2007). Chemical combination effects predict connectivity in biological systems. *Molecular Systems Biology* **3**.

Leidner, J., Palkowitsch, L., Marienfeld, U., Fischer, D. and Marienfeld, R. (2008). Identification of lysine residues critical for the transcriptional activity and polyubiquitination of the NF-Î°B family member RelB. *Biochem J* **416,** 117-127.

Leung, J. Y., Kolligs, F. T., Wu, R., Zhai, Y., Kuick, R., Hanash, S., Cho, K. R. and Fearon, E. R. (2002). Activation of AXIN2 Expression by Î2-Catenin-T Cell Factor. *Journal of Biological Chemistry* **277,** 21657-21665.

Li, Q., Lu, Q., Hwang, J. Y., BÃ¼scher, D., Lee, K.-F., Izpisua-Belmonte, J. C. and Verma, I. M. (1999). IKK1-deficient mice exhibit abnormal development of skin and skeleton *Genes & Development* **13** 1322-1328

Lin, X., Mu, Y., Cunningham, E. T., Jr., Marcu, K. B., Geleziunas, R. and Greene, W. C. (1998). Molecular Determinants of NF-kappa B-Inducing Kinase Action. *Mol. Cell. Biol.* **18,** 5899-5907.

Ling, L., Cao, Z. and Goeddel, D. V. (1998). NF-kappB-inducing kinase activates IKK-alpha by phosphorylation of Ser-176. *Proceedings of the National Academy of Sciences* **95,** 3792-3797.

Lipniacki, T., Paszek, P., Brasier, A. R., Luxon, B. and Kimmel, M. (2004). Mathematical model of NF-[kappa]B regulatory module. *Journal of Theoretical Biology* **228,** 195-215.

Liu, C. and He, X. (2010). Destruction of a destructor: a new avenue for cancer therapeutics targeting the Wnt pathway. *J Mol Cell Biol* **2,** 70-73.

Liu, C., Kato, Y., Zhang, Z., Do, V. M., Yankner, B. A. and He, X. (1999). beta -Trcp couples beta -catenin phosphorylation-degradation and regulates Xenopus axis formation. *PNAS* **96,** 6273-6278.

Liu, C., Li, Y., Semenov, M., Han, C., Baeg, G. H., Tan, Y., Zhang, Z., Lin, X. and He, X. (2002). Control of beta-catenin phosphorylation/degradation by a dual-kinase mechanism. *Cell* **108,** 837-847.

Liu, J., Stevens, J., Rote, C. A., Yost, H. J., Hu, Y., Neufeld, K. L., White, R. L. and Matsunami, N. (2001). Siah-1 mediates a novel beta-catenin degradation pathway linking p53 to the adenomatous polyposis coli protein. *Mol Cell* **7,** 927-936.

Liu, J., Xing, Y., Hinds, T. R., Zheng, J. and Xu, W. (2006). The third 20 amino acid repeat is the tightest binding site of APC for beta-catenin. *J Mol Biol* **360,** 133-144.

Llorens, M., Nuno, J. C., Rodriguez, Y., Melendez-Hevia, E. and Montero, F. (1999). Generalization of the theory of transition times in metabolic pathways: a geometrical approach. *Biophys J* **77,** 23-36.

Logan, C. Y. and Nusse, R. (2004). The Wnt signaling pathway in development and disease. *Annu Rev Cell Dev Biol* **20,** 781-810.

Lombardi, L., Ciana, P., Cappellini, C., Trecca, D., Guerrini, L., Migliazza, A., Teresa Maiolo, A. and Neri, A. (1995). Structural and functional characterization of the promoter regions of the NFKB2 gene. *Nucleic Acids Research* **23,** 2328-2336.

Luftig, M. A., Cahir-McFarland, E., Mosialos, G. and Kieff, E. (2001). Effects of the NIK aly Mutation on NF-κB Activation by the Epstein-Barr Virus Latent Infection Membrane Protein, Lymphotoxin β Receptor, and CD40. *Journal of Biological Chemistry* **276,** 14602-14606.

Luo, W., Peterson, A., Garcia, B. A., Coombs, G., Kofahl, B., Heinrich, R., Shabanowitz, J., Hunt, D. F., Yost, H. J. and Virshup, D. M. (2007). Protein phosphatase 1 regulates assembly and function of the beta-catenin degradation complex. *Embo J* **26,** 1511-1521.

Lustig, B., Jerchow, B., Sachs, M., Weiler, S., Pietsch, T., Karsten, U., van de Wetering, M., Clevers, H., Schlag, P. M., Birchmeier, W. and Behrens, J. (2002). Negative Feedback Loop of Wnt Signaling through Upregulation of Conductin/Axin2 in Colorectal and Liver Tumors. *Mol. Cell. Biol.* **22,** 1184-1193.

MacDonald, B. T., Tamai, K. and He, X. (2009). Wnt/beta-catenin signaling: components, mechanisms, and diseases. *Dev Cell* **17,** 9-26.

Mao, J., Wang, J., Liu, B., Pan, W., Farr, I., Gist H., Flynn, C., Yuan, H., Takada, S. and Kimelman, D. (2001). Low-Density Lipoprotein Receptor-Related Protein-5 Binds to Axin and Regulates the Canonical Wnt Signaling Pathway. *Molecular Cell* **7,** 801-809.

Mathes, E., O'Dea, E. L., Hoffmann, A. and Ghosh, G. (2008). NF-kappaB dictates the degradation pathway of IkappaBalpha. *Embo J* **27,** 1357-1367.

McNeill, H. and Woodgett, J. R. (2010). When pathways collide: collaboration and connivance among signalling proteins in development. *Nat Rev Mol Cell Biol* **11**, 404-413.

Mirams, G. R., Byrne, H. M. and King, J. R. (2010). A multiple timescale analysis of a mathematical model of the Wnt/beta-catenin signalling pathway. *J Math Biol* **60**, 131-160.

Miyoshi, Y., Iwao, K., Nagasawa, Y., Aihara, T., Sasaki, Y., Imaoka, S., Murata, M., Shimano, T. and Nakamura, Y. (1998). Activation of the beta-catenin gene in primary hepatocellular carcinomas by somatic alterations involving exon 3. *Cancer Res* **58**, 2524-2527.

Miyoshi, Y., Nagse, H., Ando, H., Horii, A., Ichii, S., Nakatsuru, S., Aoki, T., Miki, Y., Mori, T. and Nakamura, Y. (1992). Somatic mutations of the APC gene in colorectal tumors: mutation cluster region in the APC gene. *Human Molecular Genetics* **1**, 229-233.

Monk, N. A. (2003). Oscillatory expression of Hes1, p53, and NF-kappaB driven by transcriptional time delays. *Curr Biol* **13**, 1409-1413.

Moon, R. T., Kohn, A. D., De Ferrari, G. V. and Kaykas, A. (2004). WNT and beta-catenin signalling: diseases and therapies. *Nat Rev Genet* **5**, 691-701.

Mosimann, C., Hausmann, G. and Basler, K. (2009). Beta-catenin hits chromatin: regulation of Wnt target gene activation. *Nat Rev Mol Cell Biol* **10**, 276-286.

Müller, J. R. and Siebenlist, U. (2003). Lymphotoxin beta receptor induces sequential activation of distinct NF-kappa B factors via separate signaling pathways. *J Biol Chem* **278**, 12006-12012.

Munemitsu, S., Albert, I., Souza, B., Rubinfeld, B. and Polakis, P. (1995). Regulation of Intracellular {beta}-Catenin Levels by the Adenomatous Polyposis Coli (APC) Tumor-Suppressor Protein. *PNAS* **92**, 3046-3050.

Murray, P. J., Kang, J. W., Mirams, G. R., Shin, S. Y., Byrne, H. M., Maini, P. K. and Cho, K. H. (2010). Modelling spatially regulated beta-catenin dynamics and invasion in intestinal crypts. *Biophys J* **99**, 716-725.

Naik, S. and Piwnica-Worms, D. (2007). Real-time imaging of beta-catenin dynamics in cells and living mice. *Proceedings of the National Academy of Sciences* **104**, 17465-17470.

Nelander, S., Wang, W., Nilsson, B., She, Q.-B., Pratilas, C., Rosen, N., Gennemark, P. and Sander, C. (2008). Models from experiments: combinatorial drug perturbations of cancer cells. *Molecular Systems Biology* **4**.

Nhieu, J. T., Renard, C. A., Wei, Y., Cherqui, D., Zafrani, E. S. and Buendia, M. A. (1999). Nuclear accumulation of mutated beta-catenin in hepatocellular carcinoma is associated with increased cell proliferation. *Am J Pathol* **155**, 703-710.

Niehrs, C. and Shen, J. (2010). Regulation of Lrp6 phosphorylation. *Cellular and Molecular Life Sciences* **67**, 2551-2562.

Niida, A., Hiroko, T., Kasai, M., Furukawa, Y., Nakamura, Y., Suzuki, Y., Sugano, S. and Akiyama, T. (2004). DKK1, a negative regulator of Wnt signaling, is a target of the [beta]-catenin//TCF pathway. **23**, 8520-8526.

Nusse, R. (2003). Wnts and Hedgehogs: lipid-modified proteins and similarities in signaling mechanisms at the cell surface. *Development* **130**, 5297-5305.

Park, C. H., Chang, J. Y., Hahm, E. R., Park, S., Kim, H.-K. and Yang, C. H. (2005). Quercetin, a potent inhibitor against [beta]-catenin/Tcf signaling in SW480 colon cancer cells. *Biochemical and Biophysical Research Communications* **328**, 227-234.

Park, S., Gwak, J., Cho, M., Song, T., Won, J., Kim, D.-E., Shin, J.-G. and Oh, S. (2006). Hexachlorophene Inhibits Wnt/beta-Catenin Pathway by Promoting Siah-Mediated beta-Catenin Degradation. *Mol Pharmacol* **70**, 960-966.

Perkins, N. D. (2007). Integrating cell-signalling pathways with NF-kappaB and IKK function. *Nat Rev Mol Cell Biol* **8**, 49-62.

Polakis, P. (2000). Wnt signaling and cancer. *Genes Dev* **14**, 1837-1851.

Polakis, P. (2007). The many ways of Wnt in cancer. *Curr Opin Genet Dev* **17**, 45-51.

Press, W. h., Teukolsky, S. A., Vetterling, W. T. and Flannery, B. P. *Numerical Recipies: The Art of Science Computing*, 3 edn.

Qing, G., Qu, Z. and Xiao, G. (2005). Stabilization of basally translated NF-kappaB-inducing kinase (NIK) protein functions as a molecular switch of processing of NF-kappaB2 p100. *J Biol Chem* **280**, 40578-40582.

Qing, G. and Xiao, G. (2005). Essential Role of IkappaB Kinase alpha in the Constitutive Processing of NF-kappaB2 p100. *Journal of Biological Chemistry* **280**, 9765-9768.

Ramis-Conde, I., Drasdo, D., Anderson, A. R. and Chaplain, M. A. (2008). Modeling the influence of the E-cadherin-beta-catenin pathway in cancer cell invasion: a multiscale approach. *Biophys J* **95**, 155-165.

Razani, B., Zarnegar, B., Ytterberg, A. J., Shiba, T., Dempsey, P. W., Ware, C. F., Loo, J. A. and Cheng, G. (2010). Negative feedback in noncanonical NF-kappaB signaling modulates NIK stability through IKKalpha-mediated phosphorylation. *Sci Signal* **3**, ra41.

Riese, J., Yu, X., Munnerlyn, A., Eresh, S., Hsu, S.-C., Grosschedl, R. and Bienz, M. (1997). LEF-1, a Nuclear Factor Coordinating Signaling Inputs from wingless and decapentaplegic. *Cell* **88**, 777-787.

Rodriguez-Gonzalez, J. G., Santillan, M., Fowler, A. C. and Mackey, M. C. (2007). The segmentation clock in mice: interaction between the Wnt and Notch signalling pathways. *J Theor Biol* **248**, 37-47.

Ryseck, R. P., Bull, P., Takamiya, M., Bours, V., Siebenlist, U., Dobrzanski, P. and Bravo, R. (1992). RelB, a new Rel family transcription activator that can interact with p50-NF-kappa B. *Mol. Cell. Biol.* **12**, 674-684.

Saegusa, M., Hashimura, M., Kuwata, T., Hamano, M. and Okayasu, I. (2007). Crosstalk between NF-kappaB/p65 and beta-catenin/TCF4/p300 signalling pathways through alterations in GSK-3beta expression during <I>trans</I>-differentiation of endometrial carcinoma cells. *The Journal of Pathology* **213**, 35-45.

Salahshor, S. and Woodgett, J. R. (2005). The links between axin and carcinogenesis. *J Clin Pathol* **58**, 225-236.

Salazar, C. and Höfer, T. (2009). Multisite protein phosphorylation--from molecular mechanisms to kinetic models. *Febs J* **276**, 3177-3198.

Sanjo, H., Zajonc, D. M., Braden, R., Norris, P. S. and Ware, C. F. (2010). Allosteric regulation of the ubiquitin:NIK and TRAF3 E3 ligases by the lymphotoxin-{beta} receptor. *J Biol Chem* **285**, 17148-17155.

Satoh, S., Daigo, Y., Furukawa, Y., Kato, T., Miwa, N., Nishiwaki, T., Kawasoe, T., Ishiguro, H., Fujita, M., Tokino, T., Sasaki, Y., Imaoka, S., Murata, M., Shimano, T., Yamaoka, Y. and Nakamura, Y. (2000). AXIN1 mutations in hepatocellular carcinomas, and growth suppression in cancer cells by virus-mediated transfer of AXIN1. *Nat Genet* **24**, 245-250.

Schaber, J., Kofahl, B., Kowald, A. and Klipp, E. (2006). A modelling approach to quantify dynamic crosstalk between the pheromone and the starvation pathway in baker's yeast. *FEBS J* **273**, 3520-3533.

Scheidereit, C. (2006). IkappaB kinase complexes: gateways to NF-kappaB activation and transcription. *Oncogene* **25**, 6685-705.

Schilling, M., Maiwald, T., Hengl, S., Winter, D., Kreutz, C., Kolch, W., Lehmann, W. D., Timmer, J. and Klingmuller, U. (2009). Theoretical and experimental analysis links isoform- specific ERK signalling to cell fate decisions. *Molecular Systems Biology* **5**, 334.

Schoeberl, B., Eichler-Jonsson, C., Gilles, E. D. and Muller, G. (2002). Computational modeling of the dynamics of the MAP kinase cascade activated by surface and internalized EGF receptors. *Nat Biotechnol* **20**, 370-375.

Seaton, D. and Krishnan, J. (2011). The coupling of pathways and processes through shared components. *BMC Systems Biology* **5**, 103.

Senftleben, U., Cao, Y., Xiao, G., Greten, F. R., Krahn, G., Bonizzi, G., Chen, Y., Hu, Y., Fong, A., Sun, S.-C. and Karin, M. (2001). Activation by IKKalpha of a Second, Evolutionary Conserved, NF-kappa B Signaling Pathway. *Science* **293**, 1495-1499.

Shih, V. F., Kearns, J. D., Basak, S., Savinova, O. V., Ghosh, G. and Hoffmann, A. (2009). Kinetic control of negative feedback regulators of NF-kappaB/RelA determines their pathogen- and cytokine-receptor signaling specificity. *Proc Natl Acad Sci U S A* **106**, 9619-9624.

Shih, V. F., Tsui, R., Caldwell, A. and Hoffmann, A. (2011). A single NFkappaB system for both canonical and non-canonical signaling. *Cell Res* **21**, 86-102.

Shin, S. Y., Rath, O., Zebisch, A., Choo, S. M., Kolch, W. and Cho, K. H. (2010). Functional roles of multiple feedback loops in extracellular signal-regulated kinase and Wnt signaling pathways that regulate epithelial-mesenchymal transition. *Cancer Res* **70**, 6715-6724.

Shinkura, R., Kitada, K., Matsuda, F., Tashiro, K., Ikuta, K., Suzuki, M., Kogishi, K., Serikawa, T. and Honjo, T. (1999). Alymphoplasia is caused by a point mutation in the mouse gene encoding Nf-[kappa]b-inducing kinase. **22**, 74-77.

Sick, S., Reinker, S., Timmer, J. and Schlake, T. (2006). WNT and DKK Determine Hair Follicle Spacing Through a Reaction-Diffusion Mechanism. *Science* **314**, 1447-1450.

Silke, J. and Brink, R. (2010). Regulation of TNFRSF and innate immune signalling complexes by TRAFs and cIAPs. *Cell Death Differ* **17**, 35-45.

Spiegelman, V. S., Slaga, T. J., Pagano, M., Minamoto, T., Ronai, Z. e. and Fuchs, S. Y. (2000). Wnt/[beta]-Catenin Signaling Induces the Expression and Activity of [beta]TrCP Ubiquitin Ligase Receptor. *Molecular Cell* **5**, 877-882.

Städeli, R., Hoffmans, R. and Basler, K. (2006). Transcription under the Control of Nuclear Arm/[beta]-Catenin. *Current Biology* **16**, R378-R385.

Stahl, S., Ittrich, C., Marx-Stoelting, P., Kohle, C., Altug-Teber, O., Riess, O., Bonin, M., Jobst, J., Kaiser, S., Buchmann, A. and Schwarz, M. (2005). Genotype-phenotype relationships in hepatocellular tumors from mice and man. *Hepatology* **42**, 353-361.

Staudt, L. M. (2010). Oncogenic Activation of NF-kappaB. *Cold Spring Harbor Perspectives in Biology* **2**, a000109.

Sun, S. C. (2011). Non-canonical NF-kappaB signaling pathway. *Cell Res* **21**, 71-85.

Sun, S. C., Ganchi, P. A., Ballard, D. W. and Greene, W. C. (1993). NF-kappa B controls expression of inhibitor I kappa B alpha: evidence for an inducible autoregulatory pathway. *Science* **259**, 1912-1915.

Takada, R., Hijikata, H., Kondoh, H. and Takada, S. (2005). Analysis of combinatorial effects of Wnts and Frizzleds on -catenin/armadillo stabilization and Dishevelled phosphorylation. *Genes to Cells* **10**, 919-928.

Takigawa, Y. and Brown, A. M. (2008). Wnt signaling in liver cancer. *Curr Drug Targets* **9**, 1013-1024.

Tamai, K., Zeng, X., Liu, C., Zhang, X., Harada, Y., Chang, Z. and He, X. (2004). A mechanism for Wnt coreceptor activation. *Mol Cell* **13**, 149-156.

Taniguchi, K., Roberts, L. R., Aderca, I. N., Dong, X., Qian, C., Murphy, L. M., Nagorney, D. M., Burgart, L. J., Roche, P. C., Smith, D. I., Ross, J. A. and Liu, W. (2002). Mutational spectrum of beta-catenin, AXIN1, and AXIN2 in hepatocellular carcinomas and hepatoblastomas. *Oncogene* **21**, 4863-4871.

Tauriello, D. V., Haegebarth, A., Kuper, I., Edelmann, M. J., Henraat, M., Canninga-van Dijk, M. R., Kessler, B. M., Clevers, H. and Maurice, M. M. (2010). Loss of the tumor suppressor CYLD enhances Wnt/beta-catenin signaling through K63-linked ubiquitination of Dvl. *Mol Cell* **37**, 607-619.

Tay, S., Hughey, J. J., Lee, T. K., Lipniacki, T., Quake, S. R. and Covert, M. W. (2010). Single-cell NF-kappaB dynamics reveal digital activation and analogue information processing. *Nature* **466**, 267-271.

ten Berge, D., Koole, W., Fuerer, C., Fish, M., Eroglu, E. and Nusse, R. (2008). Wnt signaling mediates self-organization and axis formation in embryoid bodies. *Cell Stem Cell* **3**, 508-518.

Thu, Y. M. and Richmond, A. (2010). NF-kappaB inducing kinase: a key regulator in the immune system and in cancer. *Cytokine Growth Factor Rev* **21**, 213-226.

Tolwinski, N. S. and Wieschaus, E. (2004). Rethinking WNT signaling. *Trends Genet* **20**, 177-181.

Tyson, J. J., Chen, K. C. and Novak, B. (2003). Sniffers, buzzers, toggles and blinkers: dynamics of regulatory and signaling pathways in the cell. *Curr Opin Cell Biol* **15**, 221-231.

Vallabhapurapu, S., Matsuzawa, A., Zhang, W., Tseng, P. H., Keats, J. J., Wang, H., Vignali, D. A., Bergsagel, P. L. and Karin, M. (2008). Nonredundant and complementary functions of TRAF2 and TRAF3 in a ubiquitination cascade that activates NIK-dependent alternative NF-kappaB signaling. *Nat Immunol* **9**, 1364-1370.

van Amerongen, R., Mikels, A. and Nusse, R. (2008). Alternative wnt signaling is initiated by distinct receptors. *Sci Signal* **1**, re9.

van de Wetering, M., Barker, N., Harkes, I. C., van der Heyden, M., Dijk, N. J., Hollestelle, A., Klijn, J. G. M., Clevers, H. and Schutte, M. (2001). Mutant E-cadherin Breast Cancer Cells Do Not Display Constitutive Wnt Signaling *Cancer Research* **61** 278-284

van Leeuwen, I. M., Mirams, G. R., Walter, A., Fletcher, A., Murray, P., Osborne, J., Varma, S., Young, S. J., Cooper, J., Doyle, B., Pitt-Francis, J., Momtahan, L., Pathmanathan, P., Whiteley, J. P., Chapman, S. J., Gavaghan, D. J., Jensen, O. E., King, J. R., Maini, P. K., Waters, S. L. and Byrne, H. M. (2009). An integrative computational model for intestinal tissue renewal. *Cell Prolif* **42**, 617-636.

van Leeuwen, I. M. M., Byrne, H. M., Jensen, O. E. and King, J. R. (2007). Elucidating the interactions between the adhesive and transcriptional functions of [beta]-catenin in normal and cancerous cells. *Journal of Theoretical Biology* **247**, 77-102.

Vatsyayan, J., Qing, G., Xiao, G. and Hu, J. (2008). SUMO1 modification of NF-[kappa]B2/p100 is essential for stimuli-induced p100 phosphorylation and processing. **9**, 885-890.

Villacorte, M., Suzuki, K., Hayashi, K., de Sousa Lopes, S. C., Haraguchi, R., Taketo, M. M., Nakagata, N. and Yamada, G. (2010). Antagonistic crosstalk of Wnt/[beta]-catenin/Bmp signaling within the Apical Ectodermal Ridge (AER) regulates interdigit formation. *Biochemical and Biophysical Research Communications* **391**, 1653-1657.

Villanueva, A., Newell, P., Chiang, D. Y., Friedman, S. L. and Llovet, J. M. (2007). Genomics and signaling pathways in hepatocellular carcinoma. *Semin Liver Dis* **27**, 55-76.

von Dassow, G., Meir, E., Munro, E. M. and Odell, G. M. (2000). The segment polarity network is a robust developmental module. *Nature* **406**, 188-192.

Wallach, D. and Kovalenko, A. (2008). Self-termination of the terminator. *Nature Immunology* **9**, 1325-1327.

Wang, X., Adhikari, N., Li, Q., Guan, Z. and Hall, J. L. (2004). The Role of {beta}-Transducin Repeat-Containing Protein ({beta}-TrCP) in the Regulation of NF-{kappa}B in Vascular Smooth Muscle Cells. *Arterioscler Thromb Vasc Biol* **24**, 85-90.

Wawra, C., Kuhl, M. and Kestler, H. A. (2007). Extended analyses of the Wnt/beta-catenin pathway: robustness and oscillatory behaviour. *FEBS Lett* **581**, 4043-4048.

Wolf, J., Dronov, S., Tobin, F. and Goryanin, I. (2007). The impact of the regulatory design on the response of epidermal growth factor receptor-mediated signal transduction towards oncogenic mutations. *FEBS Journal* **274**, 5505-5517.

Wong, C. M., Fan, S. T. and Ng, I. O. (2001). beta-Catenin mutation and overexpression in hepatocellular carcinoma: clinicopathologic and prognostic significance. *Cancer* **92**, 136-145.

Wu, G., Huang, H., Abreu, J. G. and He, X. (2009). Inhibition of GSK3 Phosphorylation of beta-Catenin via Phosphorylated PPPSPXS Motifs of Wnt Coreceptor LRP6. *PLoS ONE* **4**, e4926.

Xiao, G., Fong, A. and Sun, S.-C. (2004). Induction of p100 Processing by NF-{kappa}B-inducing Kinase Involves Docking I{kappa}B Kinase {alpha} (IKK{alpha}) to p100 and IKK{alpha}-mediated Phosphorylation. *J. Biol. Chem.* **279**, 30099-30105.

Xiao, G., Harhaj, E. W. and Sun, S.-C. (2001). NF-[kappa]B-Inducing Kinase Regulates the Processing of NF-[kappa]B2 p100. *Molecular Cell* **7**, 401-409.

Xing, Y., Clements, W. K., Kimelman, D. and Xu, W. (2003). Crystal structure of a {beta}-catenin/Axin complex suggests a mechanism for the {beta}-catenin destruction complex. *Genes Dev.* **17**, 2753-2764.

Yanagawa, S., van Leeuwen, F., Wodarz, A., Klingensmith, J. and Nusse, R. (1995). The dishevelled protein is modified by wingless signaling in Drosophila. *Genes Dev* **9**, 1087-1097.

Yang, J., Zhang, W., Evans, P. M., Chen, X., He, X. and Liu, C. (2006). Adenomatous polyposis coli (APC) differentially regulates beta-catenin phosphorylation and ubiquitination in colon cancer cells. *J Biol Chem* **281**, 17751-17757.

Yilmaz, Z. B., Weih, D. S., Sivakumar, V. and Weih, F. (2003). RelB is required for Peyer's patch development: differential regulation of p52-RelB by lymphotoxin and TNF. *Embo J* **22**, 121-130.

Zarnegar, B. J., Wang, Y., Mahoney, D. J., Dempsey, P. W., Cheung, H. H., He, J., Shiba, T., Yang, X., Yeh, W. C., Mak, T. W., Korneluk, R. G. and Cheng, G. (2008).

Noncanonical NF-kappaB activation requires coordinated assembly of a regulatory complex of the adaptors cIAP1, cIAP2, TRAF2 and TRAF3 and the kinase NIK. *Nat Immunol* **9**, 1371-1378.

Zeng, X., Huang, H., Tamai, K., Zhang, X., Harada, Y., Yokota, C., Almeida, K., Wang, J., Doble, B., Woodgett, J., Wynshaw-Boris, A., Hsieh, J.-C. and He, X. (2008). Initiation of Wnt signaling: control of Wnt coreceptor Lrp6 phosphorylation/activation via frizzled, dishevelled and axin functions. *Development* **135**, 367-375.

Zeng, X., Tamai, K., Doble, B., Li, S., Huang, H., Habas, R., Okamura, H., Woodgett, J. and He, X. (2005). A dual-kinase mechanism for Wnt co-receptor phosphorylation and activation. *Nature* **438**, 873-877.

Zhang, Y., Liu, S., Mickanin, C., Feng, Y., Charlat, O., Michaud, G. A., Schirle, M., Shi, X., Hild, M., Bauer, A., Myer, V. E., Finan, P. M., Porter, J. A., Huang, S.-M. A. and Cong, F. (2011). RNF146 is a poly(ADP-ribose)-directed E3 ligase that regulates axin degradation and Wnt signalling. **13**, 623-629.

Zucman-Rossi, J., Benhamouche, S., Godard, C., Boyault, S., Grimber, G., Balabaud, C., Cunha, A. S., Bioulac-Sage, P. and Perret, C. (2007). Differential effects of inactivated Axin1 and activated beta-catenin mutations in human hepatocellular carcinomas. *Oncogene* **26**, 774-780.

Danksagung

Mein Dank gilt einer großen Anzahl Menschen, die diese Arbeit überhaupt ermöglicht, auf unterschiedlichste Arten beeinflusst und zu ihrem Gelingen beigetragen haben.

Zuerst möchte ich Reinhart Heinrich danken, der mir 2005 die Chance gab, in seiner Arbeitsgruppe die Promotion zum Wnt/β-Catenin-Signalweg zu beginnen. Leider konnte er die Arbeit nur bis 2006 betreuen, gab ihr jedoch wesentliche Impulse.

Edda Klipp hat sich bereit erklärt, meine universitäre Betreuung zu übernehmen. Dafür danke ich dir sehr. Außerdem möchte ich an dieser Stelle noch einmal für die Zeit danken, die ich während meiner Diplomarbeit in deiner Arbeitsgruppe verbringen durfte und die einen großen Anteil daran hatte, dass ich schließlich mit dieser Doktorarbeit angefangen habe.

Ein besonders großer und herzlicher Dank gilt Jana Wolf, die 2006 meine Betreuung übernommen hat und mir seither immer wieder mit Rat und Tat zur Seite stand. Stets fanden wir Zeit für die vielfältigsten Diskussionen. Dabei hast du nie vergessen, den Spaß an der Wissenschaft zu vermitteln. Neben dem Dank für eine großartige Betreuung auch vielen Dank für all die Gespräche und Unternehmungen abseits der Wissenschaft.

Weiter möchte ich mich sehr bei meinen Kooperationspartnern Claus Scheidereit und Buket Yilmaz, Andreas Hecht, Frank Götschel, Monika Schrempp und Katja Bruser sowie Rolf Gebhardt bedanken, die mir in vielen Diskussionen die experimentellen Aspekte der Signalwege näher gebracht haben. Vielen Dank, dass ich Teile eurer Daten in dieser Arbeit verwenden durfte. Claus und Buket, mit eurer Begeisterung für den nicht-kanonischen NF-κB-Signalweg und unser gemeinsames experimentell-theoretisches Projekt habt ihr einen großen Anteil am NF-κB-Kapitel der Arbeit. Andreas, vielen Dank für die vielen Wnt-Diskussionen und dass du all die Experimente ermöglicht hast.

Thomas Höfer und Nils Blüthgen danke ich für ihre sofortige Bereitschaft diese Arbeit zu begutachten.

Einen großen Dank an die ehemaligen und gegenwärtigen Mitglieder der Arbeitsgruppe Wolf am MDC, die stets für eine nette Arbeitsatmosphäre gesorgt haben. Bei Dorothea Busse möchte ich mich ganz herzlich für all die hilfreichen Anmerkungen zur Doktorarbeit bedanken. Ohne dich sähe sie erheblich anders aus. Antonio Politi danke ich für seine Einführung ins Datenfitten und dass er auch immer neue Nachfragen geduldig beantwortet hat. Uwe Benary möchte ich für unzählige Diskussionen und Literaturhinweise danken. Janina Mothes und Katharina Baum danke ich für spannende Diskussionen zum NF-κB-

Signalweg und Fragen zur Robustheit sowie exzellente Ablenk-Runden. Ein Dank an Dinto für die Diskussionen zum kanonischen NF-κB-Signalweg.

Vielen Dank an alle damaligen Mitglieder der Arbeitsgruppen Heinrich und Höfer für eine tolle Zeit in der Theoretischen Biophysik an der HU. Ein großes Dankeschön an Jana Schütze, die jahrelang meine Tischnachbarin war, unzählige Tassen Tee mit mir getrunken hat und trotz anderem Arbeitsgebiet immer ein offenes Ohr für all meine Signalwegsanliegen hatte. Ebenfalls vielen Dank für dein aufmerksames Korrekturlesen der Arbeit. Auch Roland Krüger möchte ich an dieser Stelle danken. Er hat vor mir am Wnt-Signalweg gearbeitet und besonders in der Anfangszeit geholfen, den Signalweg und das Modell besser zu verstehen.

Jens Timmer gilt mein Dank dafür, dass ich vorübergehend Mitglied seiner Arbeitsgruppe sein durfte.

Dirk, Claudia und Lukas danke ich für viele nette Abende und Ablenkungen sowie Korrekturen und Anmerkungen und Sofort-Hilfen bei allen Fragen zu Word.

Schließlich möchte ich meinen Eltern für ihre uneingeschränkte Unterstützung danken. Mein Dank gilt ebenfalls Gunnar, Julia, Jannis und Lennard.